An Introduction to Project Planning

An Introduction to Project Planning

Second Edition

by Jack Gido

Industrial Press Inc.

200 Madison Avenue, New York, NY 10157

To my wife, Rosemary, for her love, support and understanding.

Library of Congress Cataloging in Publication Data

Gido, Jack, 1945–
An introduction to project planning.

Bibliography: p.
Includes index.
1. Network analysis (Planning) 2. Industrial
project management. I. Title.
T57.85.G49 1985 658.4'04 85-168
ISBN 0-8311-1160–7

Second Edition

Contents

Preface

An Introduction to Project Planning, 2nd edition, is an easy-to-understand book written for everyone involved in projects. It presents a clear and concise explanation of the fundamental concepts of project planning, scheduling, and control, and illustrates how these processes are applied by following through an actual example. It is appropriate for all types and sizes of projects.

The need for a text of this nature has arisen from the following conditions:

1. Until now, most books that have been written about network planning techniques, such as PERT and CPM, have been technically and mathematically oriented. They have been filled with rather complex mathematical expressions and notations and have used technical examples. Other books have presented network planning concepts only, but have not shown how the concepts are applied.

2. There is a lack of use of network planning techniques for some of the following reasons:

   (a) Some people think that network planning is an overly sophisticated and complex technique, since it usually has been associated with projects for NASA and the Department of Defense. Therefore, there is a fear of its being overburdening for use on smaller, less complex projects.

   (b) In some instances the basic concepts and benefits of network planning are not fully understood.

As a result, although these techniques have been used effectively to control various types of projects, many companies and individuals have never attempted to use network planning techniques.

3. In some cases, network planning techniques have been misused or misapplied. Some users have merely paid lip service to network planning techniques, and some contractors have been forced to use them only because of the contract requirements of government projects. But worst of all, some people have misused these techniques by trying to use them for purposes other than planning, scheduling, and controlling project activity, such as using them only for documenting rather than for controlling the progress of a project.

An Introduction to Project Planning, 2nd ed., is written for people of all backgrounds. Network planning can be applied in many different areas including engineering, manufacturing, education, and information systems. It is especially helpful for projects that require an interaction of several functions or organizations.

There are many benefits and positive results from the implementation of network planning, such as the following:

1. Provides a master plan for a project.
2. Shows interrelationships of all portions of a project.
3. Forces the user to "think through" the entire project completely at the beginning of the project.
4. Allows the user to plan his or her own method of accomplishing the project.
5. Takes uncertainties into account.
6. Allows the user to simulate alternate plans or courses of action during both the initial development of a plan and any replanning throughout the life of the project.
7. Provides a vehicle for reporting on actual, expected, and allowable progress and also on the project status by supplying information on completed, in-progress, and unstarted tasks.
8. Forecasts when each task is expected to start and finish, and

when it must start and finish to achieve the project objectives on time.

9. Points out the areas of a project that are behind schedule or may be in potential trouble, and thus allows for "management by exception."
10. Provides a quantitative basis for making decisions on replanning and controlling a project to alleviate bottlenecks and critical areas, and thus replaces "seat of the pants" decision making.
11. Helps in planning resource requirements, allocation, and usage at the beginning and throughout a project.
12. Provides a communication link for a smoother flow of information, which results in better coordination among all people involved.

An Introduction to Project Planning contains a minimum of technical terms, and it uses primarily the network-related terminology that is acquired as the text is read. Furthermore, the mathematics are kept quite simple, and the examples are ones that can be easily understood. The chapters are designed to incorporate both the *understanding* and *use* of network planning by first presenting the concepts, and then showing how these concepts are used by applying them to an example.

Chapter 1 is an introductory chapter. It covers how network planning was developed and compares it with the more familiar bar chart technique. There also are brief discussions of the planning, scheduling, and controlling functions. Finally, it discusses when network planning can be used on a project.

Chapters 2, 3, and 5 elaborate, respectively, on planning, scheduling, and controlling a project.

Chapter 2 discusses the fundamentals of using a network diagram to plan a project. It begins by covering some initial steps necessary to start the planning function. This is followed by an explanation of the basic concepts of network planning and of the fundamental principles and building blocks used in constructing a network diagram. Finally discussed is how to prepare a network diagram for a project. The project plan, in the form of a network diagram, is the end result of the planning function.

Chapter 3 elaborates on the basic concepts of developing a schedule for the network plan constructed in Chapter 2. It begins with a discussion of the data needed to produce a schedule. It tells what data are needed, how to obtain them, when to use them, and their implications. The middle portion of this chapter gives the basic methods for calculating a schedule. The schedule is a timetable for the plan; it tells when various tasks can be expected to start and finish and also when they must start and finish in order to complete the project objectives on time. The final part of the chapter explains how to analyze the schedule to determine the status of the project.

Chapter 5 is a discussion of how to control the progress of a project. It begins with an explanation of the effects that past and current accomplishments will have on the remaining portion of the project. Covered next is how to analyze the network plan and schedule to determine the critical areas of the project and the causes of any change in the schedule. Following is a discussion of replanning to alleviate any critical or problem areas. Finally, there is an explanation of updating the schedule to take past accomplishments and plan revisions into account.

Chapter 4 explains some basic probability concepts, and shows how they may be used in conjunction with the project schedule. There also is an explanation of how to determine the probability of completing the entire project or portions of the project before their scheduled or required time.

In Chapters 2 through 5 the reader is periodically called upon to respond to review questions on recently covered material. These questions enable the reader to check his or her understanding of the material. The reader should attempt to answer these questions either mentally, or better, using a sheet of paper on which to write out the answers. The ease or difficulty with which the reader can answer these questions will help him or her decide what pace is necessary to learn and comprehend the contents thoroughly. The correct answers to these questions can be checked in Appendix G in the back of the book.

These four chapters develop an example of the application of the concepts contained in these chapters. This example, which is the project of building an outhouse, ties together the concepts in these chapters, and permits the reader to follow through a single project and see how all the concepts are applied. Moreover, the network diagrams for the outhouse example are contained in the appendices at the back of the text. The student is advised to paper clip these appendices for ready reference at the times specified in the text.

Finally, each of these four chapters has a quiz at the end so that the reader can check his or her comprehension of the principles and concepts covered. The answers to these questions are found in Appendix H in the back of the book.

Chapter 6 discusses resource considerations when using network planning.

Chapter 7 covers the role of the computer in assisting with the planning and scheduling. The beginning of the chapter discusses what factors should be considered in deciding whether or not to use a computer, and the remaining portion of the chapter lists and explains various features of computer programs used for project planning and scheduling. Appendix L provides an alphabetical listing of companies that supply computer programs for project planning, scheduling, and control.

Chapter 8 is a summary and conclusion that discusses the advantages and difficulties of using network planning.

Chapter 9 is an illustration of how network planning was implemented on a project. It describes how the initial plan was developed, how the data that were necessary to calculate a schedule were gathered, how the schedule was produced, and the methods used to control the project progress.

A glossary of terms used throughout the text is provided in Appendix J. Also, Appendix K provides a list of references for those readers who wish to pursue project management techniques further.

It should be noted that although this book illustrates the "activity-on-the-arrow" approach of network planning, the fundamental concepts can be applied to most other approaches to drawing network plans, since they all use a network diagram as a basis for developing a project plan. Chapters 2, 3, and 5 conclude by illustrating the alternate "activity-in-the-box" format.

An Introduction to Project Planning

CHAPTER 1

Development of Network Planning Techniques

Network planning is a management tool used for planning, scheduling, and controlling projects that consist of many interrelated tasks or activities.

In 1958, the United States Navy Special Projects Office, with consultants from Booz-Allen and Hamilton and with the assistance of Lockheed Missile and Space Division, developed the PERT (Program Evaluation and Review Technique) method of planning and scheduling for use on the Navy's Polaris submarine missile system.

The Navy wanted a planning technique that would take into account not only the interrelationships among the 3800 contractors, suppliers, and government agencies involved in the development and production of Polaris, but also the interdependencies of the 60,000 definable tasks necessary to complete the project.

The technique had to be able to show what effect a slippage by any contractor or a delay in starting or finishing any task would have on the other tasks as well as on the scheduled completion date for the project. The technique also had to take into account the uncertainty involved in estimating the duration of research and development tasks.

PERT met these requirements by making use of a network diagram to show the logical precedence relationships among activities and by making use of probabilistic time estimates for the expected duration of the activities. With the use of PERT, the Polaris project was completed two years ahead of its originally scheduled completion date.

About 1957 another technique, CPM (Critical Path Method), was developed by the DuPont Company in conjunction with Remington

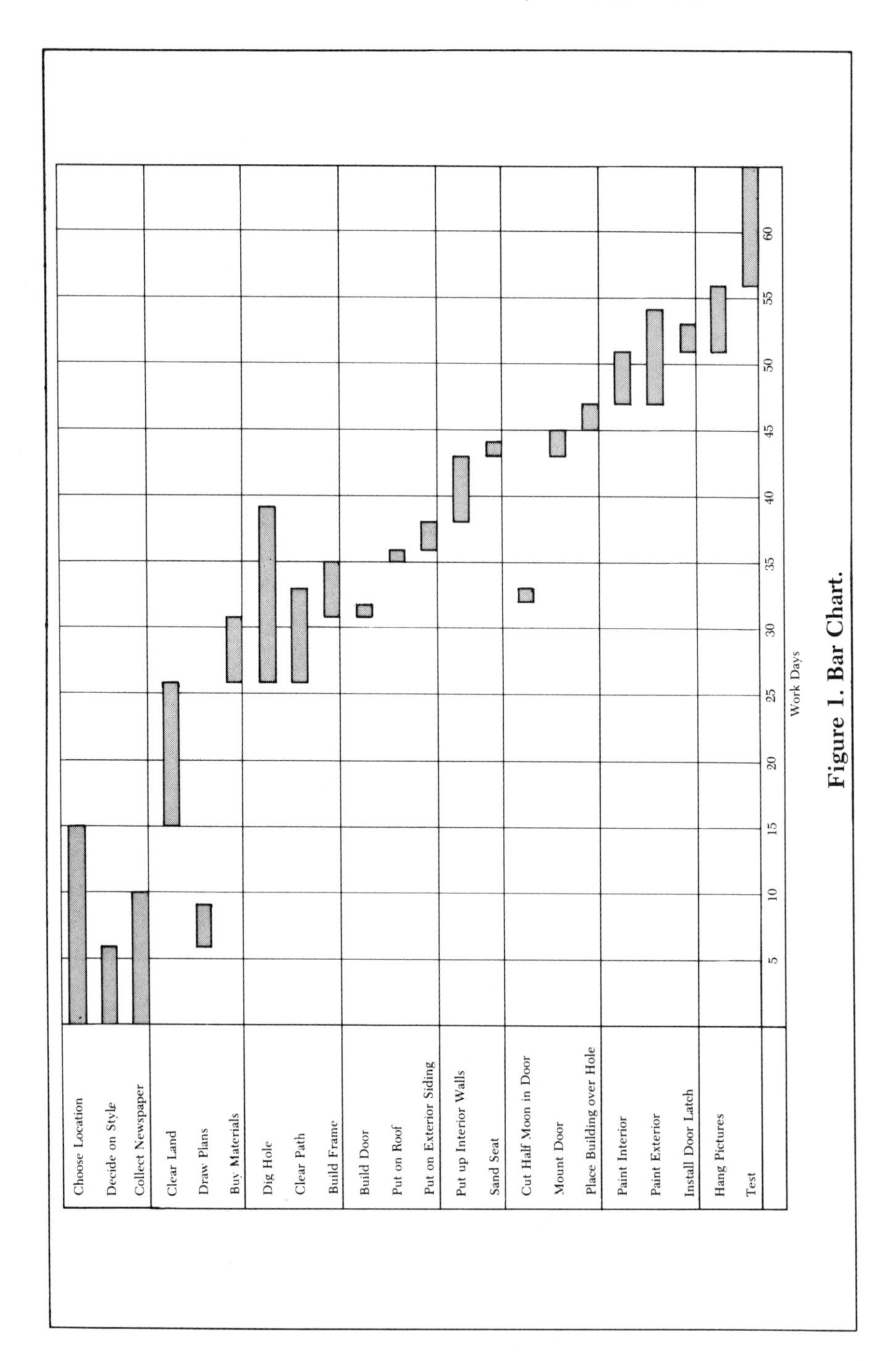

Figure 1. Bar Chart.

Rand for the purpose of optimizing the time/cost tradeoffs when undertaking routine plant overhauls and maintenance shutdowns.

Since the development of PERT and CPM, other similar techniques have also been developed, such as GERT (Graphical Evaluation and Review Technique) and PDM (Precedence Diagramming Method). All of these methods fall under the general category of Network Planning Techniques, since they all make use of a network diagram for planning.

Note: At this point, briefly scan the diagrams shown in Appendices A–F to familiarize yourself with the concept of a network diagram. These diagrams will be used later in the text as we work through the example of building an outhouse.

Network planning techniques are most often compared with the more familiar bar chart method. The network technique has several advantages over the bar chart. The primary advantage is that network techniques show interrelationships among the activities needed to complete a project, whereas bar charts do not show these interrelationships. This is due primarily to the difference in the graphical or visual portrayal of the plan used by each method. The bar chart shown in Figure 1 does not show directly that in order to buy materials both the land must be cleared and the plans must be drawn. Neither does the bar chart show what effect a delay in building the frame will have on subsequent tasks.

Another advantage in using network techniques is that the planning and scheduling functions can be separated. That is, first the project plan is developed using the network diagram, and then this plan can be scheduled afterward. However, when using a bar chart, the scheduling is done simultaneously with the planning, since the bar chart is drawn to a time scale along the horizontal axis, whereas the network diagram is not drawn to any time scale.

Plan, Schedule, and Control

Planning, scheduling, and controlling can be considered separate but interdependent functions.

Planning is the process of setting forth one's ideas about how to accomplish a certain set of objectives. Planning is the logical arrangement of the activities required to accomplish the project objective. This logical arrangement of activities is accomplished by developing a network diagram.

The network diagram is the end result of the planning function, but several steps must precede the start of the network diagram or plan. First the project objectives must be determined, then a list is made of all activities necessary to accomplish the project objectives, and finally these activities must be arranged in the form of a network diagram according to certain network principles or rules. One of the most important benefits resulting from the preparation of a network diagram is that it forces the user to develop a master plan on a systematic and logical basis.

Scheduling requires the development of a timetable to meet certain objectives for a certain plan. This timetable consists of estimates for the duration of the activities in the network diagram. Based on these estimated durations for the activities, a schedule is developed to show how long the project will take and when each activity may start and finish.

The controlling function of project planning is the process of regulating or directing the project by periodically comparing actual progress to scheduled progress. If actual progress is falling behind scheduled progress, an analysis is made to determine the cause of the deviation. Based on this analysis, certain management decisions can be made to decide on the best course of action to bring the project back on schedule. The result may be a management decision to alter the original plan and develop a revised schedule.

When to Use Network Planning

Network planning is primarily and most effectively used for projects that are somewhat unique and nonrepetitive in nature. Network planning has successfully been used on projects such as the introduction of a new product, implementation of computer systems, construction or modernization of industrial facilities, research, political campaigns, Olympics, and surgery. It is used in many functions and disciplines, including engineering, manufacturing, education, and information systems. It is especially helpful for projects that require an interaction of several functions.

Conclusion

Network planning is a management technique used for planning, scheduling, and controlling projects. Its advantages over the bar chart method become most apparent in projects where there are many interrelated tasks.

Network planning allows for the separation of planning and scheduling functions. The result of the planning function is a network diagram from which a schedule is developed. The project can be controlled by comparing actual progress with the scheduled progress.

Network planning is merely a management tool; it will not do the planning for you or make management decisions for you and it will not directly tell you how to accomplish a project. But it will help you plan your own method of accomplishing project objectives. Network planning does not replace the management decision making process; it helps the manager make decisions by indicating areas where management attention is needed.

Implementation of network planning provides many benefits, such as:

1. Provides a master plan for a project.
2. Shows interrelationships of all portions of a project.
3. Forces the user to completely "think through" the entire project at the beginning of the project.
4. Allows the user to plan his or her own method of accomplishing the project.
5. Takes uncertainties into account.
6. Allows the user to simulate alternate plans or courses of action when developing an initial plan and when replanning throughout the life of the project.
7. Provides a vehicle for reporting on actual, expected, and allowable progress and also on the project status by supplying information on completed, in-progress, and unstarted tasks.
8. Forecasts when each task is expected to start and finish, and when it must start and finish in order to achieve the project objectives on time.
9. Points out the areas of a project that are behind schedule or may be in potential trouble, and thus allows for "management by exception."
10. Provides a quantitative basis for making decisions on replanning and controlling a project to alleviate bottlenecks and critical areas, and thus replaces "seat of the pants" decision making.

11. Helps in planning resource requirements, allocation, and usage at the beginning and throughout a project.
12. Provides a communication link for a smoother flow of information, which results in better coordination among all people involved.

Planning is the process of setting forth one's ideas about how to accomplish a certain set of objectives. The planning function involves developing a network diagram. The planning process begins with defining the project objectives; next a list of all activities necessary to accomplish the project objectives is made. Then, a network diagram is drawn to portray graphically the logical precedence relationships among the activities necessary to accomplish the project objectives.

QUESTION 1. What are the three basic steps involved in the planning process?

Objectives and Activities

The first step of the planning process is to define the project objectives. The project may have a set of objectives rather than a single objective, but in any case, the objectives should be defined as clearly as possible so that it will be obvious when the objectives have been achieved. For example, the objective of a project may be the completion of a complex surgical operation or the completion of relocating a factory from one city to another. The building of an outhouse will be used as our example throughout this text, so the project objective will be the completion of the construction of the outhouse. The project objective should be specific and measurable, so it is clear and obvious when it has been accomplished.

After the project objectives have been established, the second step is to make a list of all activities necessary to accomplish the objectives. The larger and more complex a project is, the more difficult it will be to complete this list of activities. In our example, the activities necessary to build an outhouse are:

Decide on style.
Test wind direction and choose location.
Draw plans.
Buy materials.
Clear land.
Clear path from main house.
Dig hole.
Build frame.
Put on roof.
Put on exterior siding.
Put up interior walls.
Build door.
Cut half moon in door for lighting.
Mount door.
Install door latch.
Sand seat.
Position structure over hole.
Paint exterior.
Paint interior.
Hang pictures.
Collect old newspapers (25 pounds per person).
Test.

The list of activities can be developed by the project team by brainstorming techniques or by using a more structured approach known as a Work Breakdown Structure (WBS). The WBS is a graphic technique that divides and subdivides the project into phases, functions, areas, etc., similar to an organization chart. Once a WBS is prepared to the desired level of detail, then a list of activities can be developed for each bottom-most box on the chart. A possible WBS for the outhouse example is shown in Figure 2.

Once the initial steps of defining the project objectives and listing the activities necessary to accomplish those objectives have been completed, the next step is to develop a network diagram.

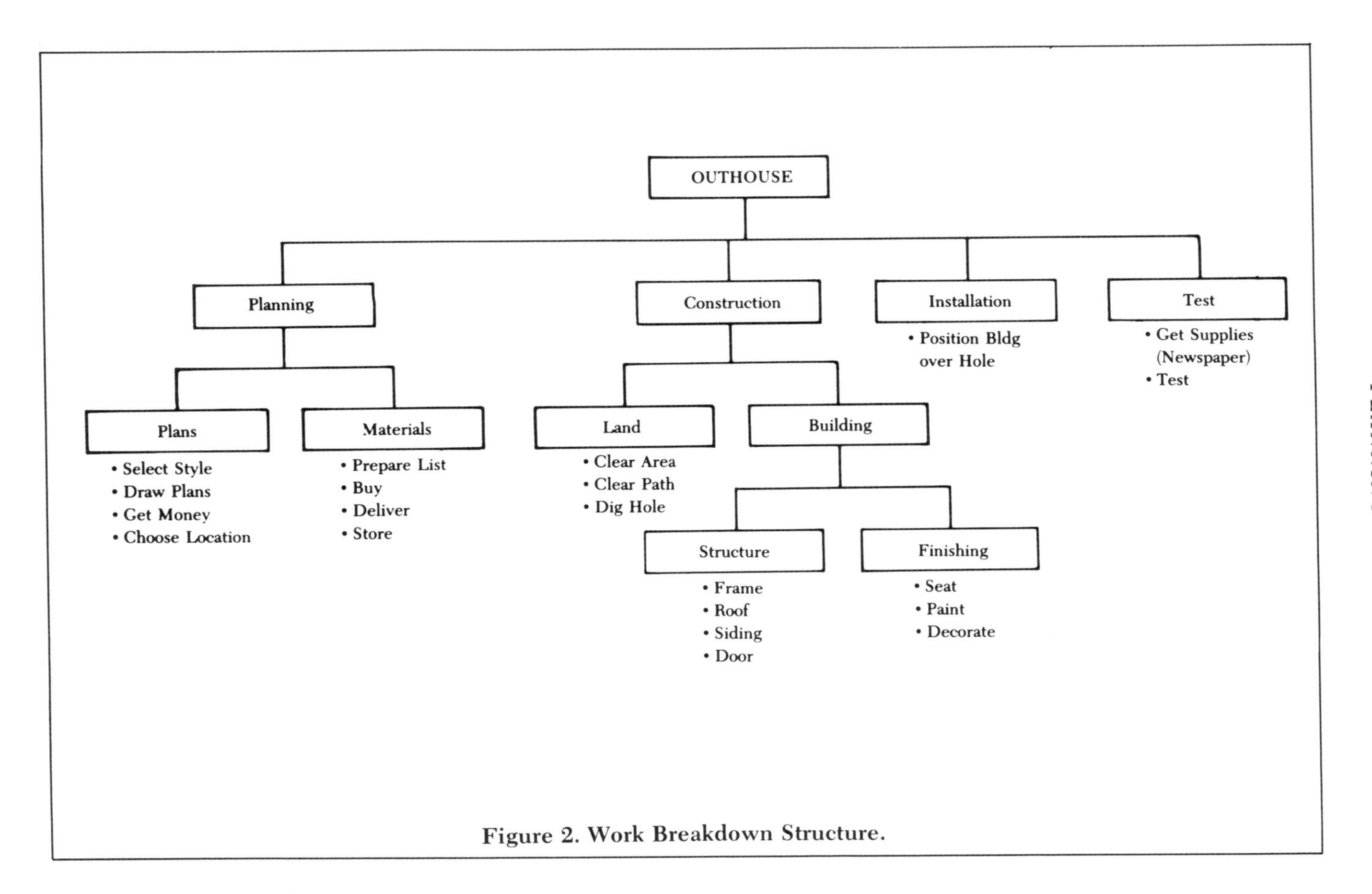

Figure 2. Work Breakdown Structure.

Network Principles

Certain basic principles must be understood when drawing a network plan.

Definitions

An **Activity** is defined as the expenditure of effort over some period of time on a particular part of a project. An activity is represented by an arrow (see Figure 3) on the network diagram. Each activity may be represented by one, and only one, arrow. The tail of the arrow designates the start of the activity, and the head of the arrow represents the completion of the activity. The length and slope of the arrow are in no way indicative of activity duration or importance (unlike the bar chart technique, in which the length of a bar indicates the duration of the activity).

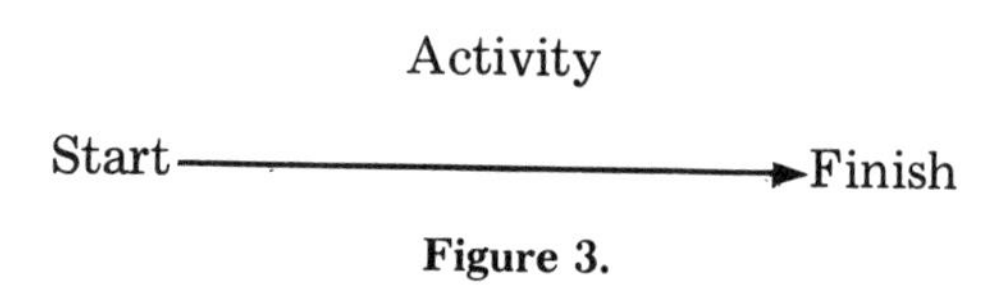

Figure 3.

An **Event** is the start or finish of an activity or group of activities. Events link activities together. An event does not consume time and is therefore considered to be an instant or point in time. Events are represented by circles (see Figure 4) on the network diagram.

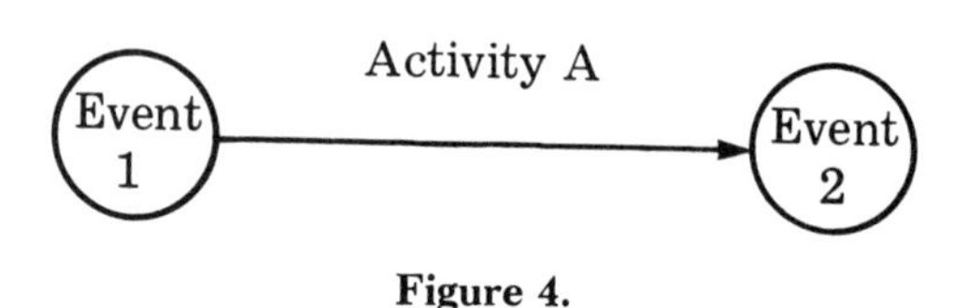

Figure 4.

The event at the start of the activity is called the activity's **predecessor event** (or start event or beginning event), and the event at the completion of the activity is called the activity's **successor event** (or finish event or ending event). For example, in Figure 4 event 1 is the start event of activity A, and event 2 is the finish event for activity A.

QUESTION 2. What is the difference between an activity and an event?

Precedence Relationships

Figure 5 shows the basic network rule of the **Precedence Relationship** of one activity to another. This figure shows that activity B cannot be started until activity A has been completed.

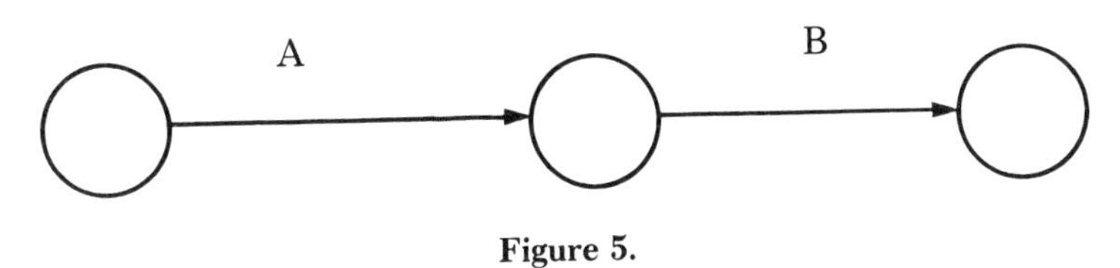

Figure 5.

Figure 5 also shows that activities may be drawn in a **series** relationship.

On the other hand, Figures 6, 7, and 8 show that activities may also be drawn in a **parallel** relationship.

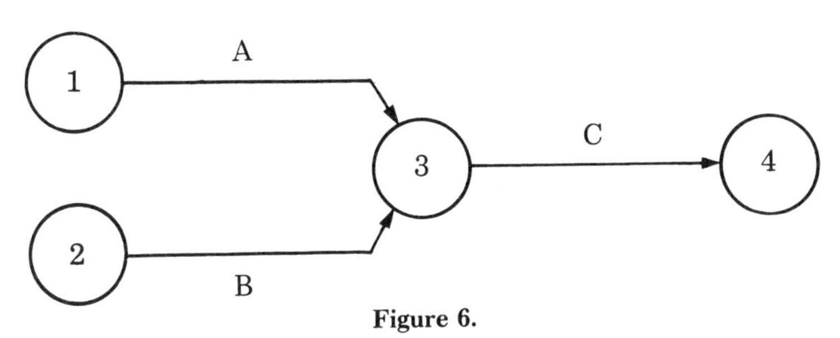

Figure 6.

Figure 6 shows that activities A and B can be performed in parallel and are therefore somewhat independent of one another. This figure also shows the precedence relationship rule: activity C cannot be started until its predecessor activities, A and B, are complete. Event 3, which represents the completion of activities A and B and also the start of activity C, can be compared to a logical "AND" function, since this event tells us that both activity A *and* activity B must be completed before activity C can start. Also, it can be said that an event has occurred when all the activities leading into that event have been completed. In Figure 6, event 3 will have occurred when activities

A and B are completed, and event 4 will have occurred when activity C is finished.

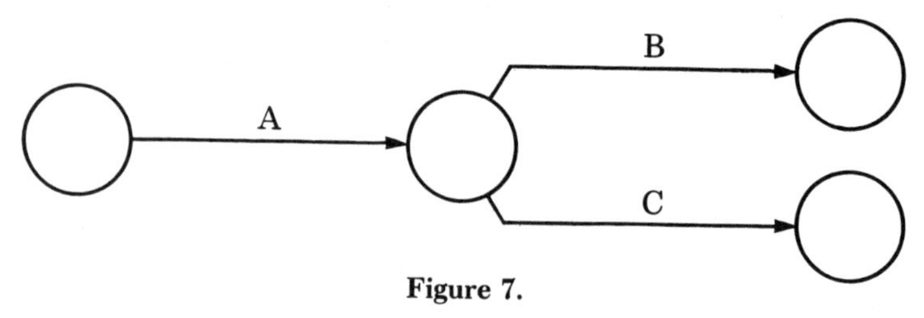

Figure 7.

Figure 7 shows that activities B and C can be performed in parallel, and it also shows that activity A must be completed before activities B and C can start.

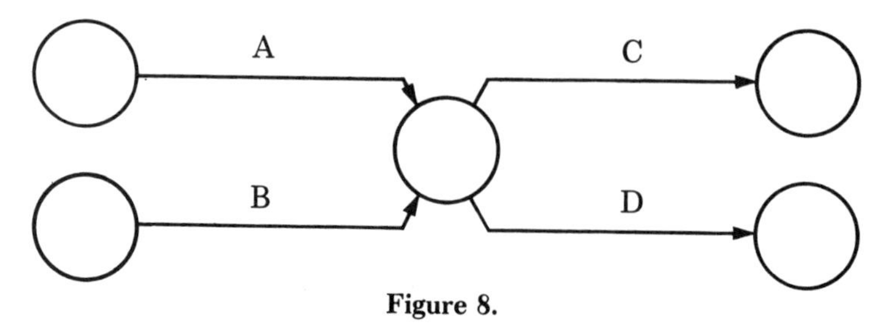

Figure 8.

Figure 8 shows that activities A and B can be performed in parallel and that when *both* A and B are completed, then activities C and D can start and be performed in parallel.

Loops

Figure 9 shows an illogical precedence relationship among the activities known as a **Loop**. These loops are not allowed when preparing a network diagram, since they would portray a path of activities that perpetually repeats itself or goes in a continuous circle.

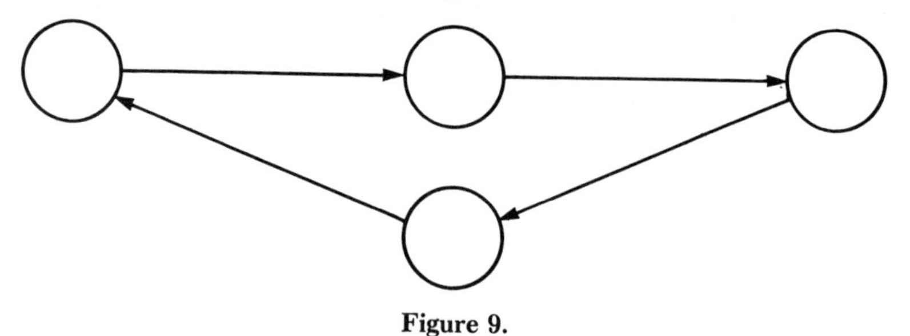

Figure 9.

Dummy Activities

There is a special type of activity known as a **Dummy** activity, which consumes zero time and is represented by a dashed arrow in the network diagram. Dummy activities are used to help in the unique identification of activities and to show logical relationships that cannot otherwise be shown.

Regarding the unique identification of activities, there are two basic rules of network diagramming:

1. Each event in the network must have a unique event number; that is, no two or more events in the network diagram can have the same event number.
2. Each activity must have a unique predecessor–successor event number combination.

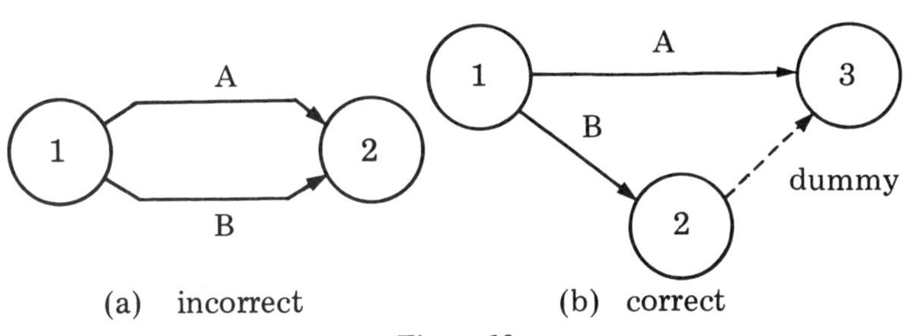

Figure 10.

For example, Figure 10a shows that both activities A and B have a predecessor–successor event number combination of 1-2. This is not allowed when preparing a network diagram, because if a person referred to activity 1-2, you would not know if he or she were talking about activity A or activity B. The addition or insertion of a dummy activity can help in identifying each activity by a unique predecessor–successor event number combination. The use of a dummy activity for this purpose is shown in Figure 10b, where activity A has a predecessor–successor event combination of 1-3, and activity B can be referred to as 1-2, and the dummy activity as 2-3.

The second use of dummy activities is to improve the logical precedence relationships among the activities in the network. For example, consider the case of four activities, A, B, C, and D. Activities A and B can be done in parallel. Activity C can start when activity A is finished, but activity D can start only when both activities A and

B are finished. To show these relationships, a dummy activity must be used as shown in Figure 11.

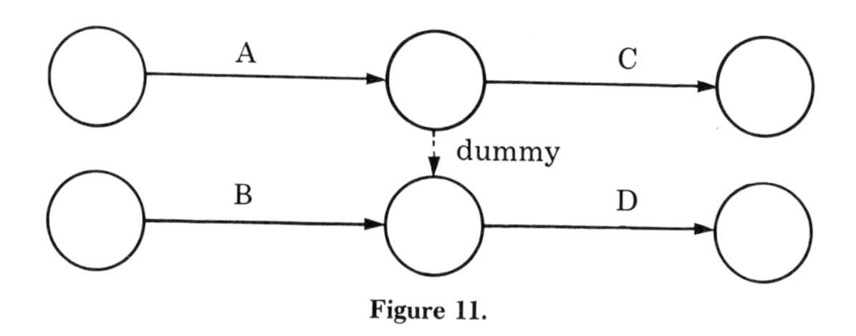

Figure 11.

Dummy activities are used to ensure that each activity in the network diagram can be identified by a unique predecessor–successor event number combination, and to help show logical precedence relationships that cannot otherwise be shown.

Laddering

Some projects have a set of activities that are repeated several consecutive times. For example, consider the project of digging a trench, laying a pipe, and backfilling for a distance of 300 feet. Assume that the project will be done in three sections of 100 feet each and that three workcrews are available, one crew each for digging the trench, laying the pipe, and backfilling. One may be inclined to draw a network diagram for this project as shown in Figure 12 or 13, but Figure 12 has all the activities in series, which means that at any one time only one crew is working while two crews are standing around. On the other hand, Figure 13 goes to the opposite extreme by showing that all three sections can be done in parallel, which is impossible, since only one crew is available for each type of activity.

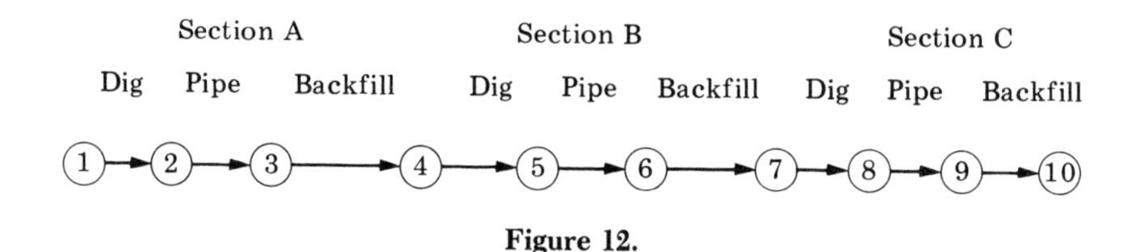

Figure 12.

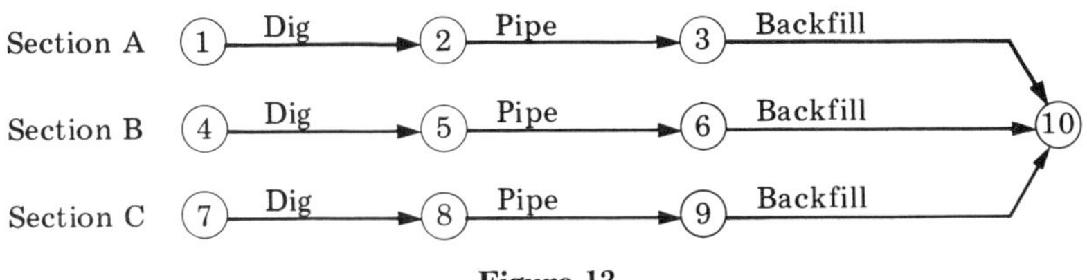

Figure 13.

Although Figure 14 shows the three sections as being somewhat in parallel, the three parallel paths are interdependent through the use of dummy activities. These dummy activities allow the project to be completed in the shortest possible time while making best use of the available resources. Figure 14 illustrates a technique known as **Laddering**.

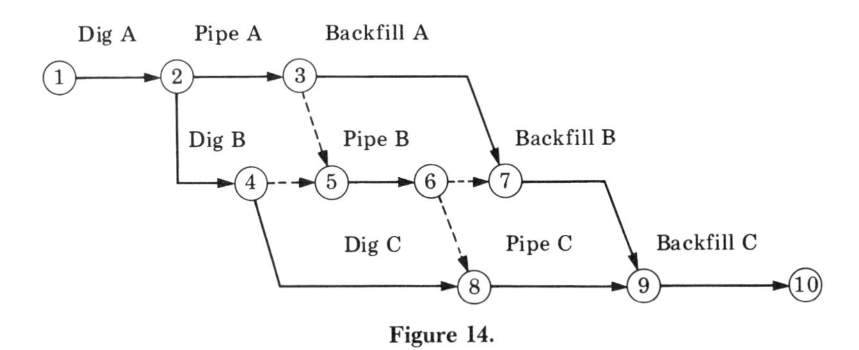

Figure 14.

Baselines

In cases where many activity arrows may have the same beginning or ending event (see Figures 15a and b), it may become messy to draw all the arrows from or to the same event. The **Baseline** technique may be used to make the network diagram a little neater. This technique allows all the arrows to or from an event to have one common line going to or coming from that event. The baseline technique is illustrated in Figures 16a and b for Figures 15a and b, respectively.

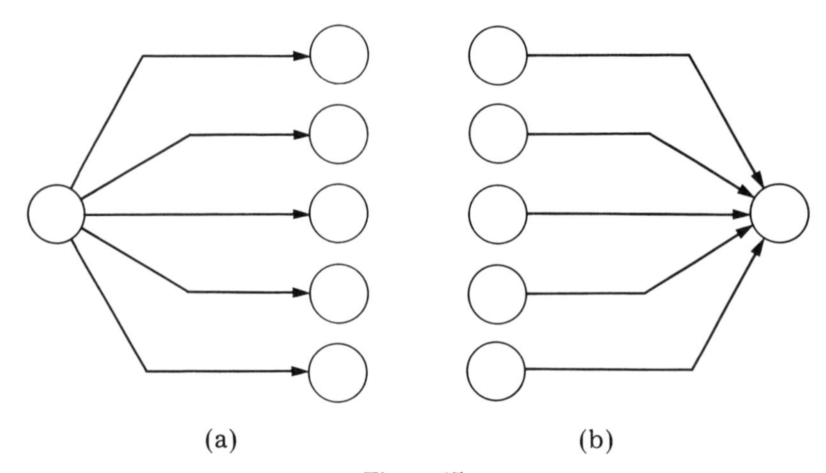

Figure 15.

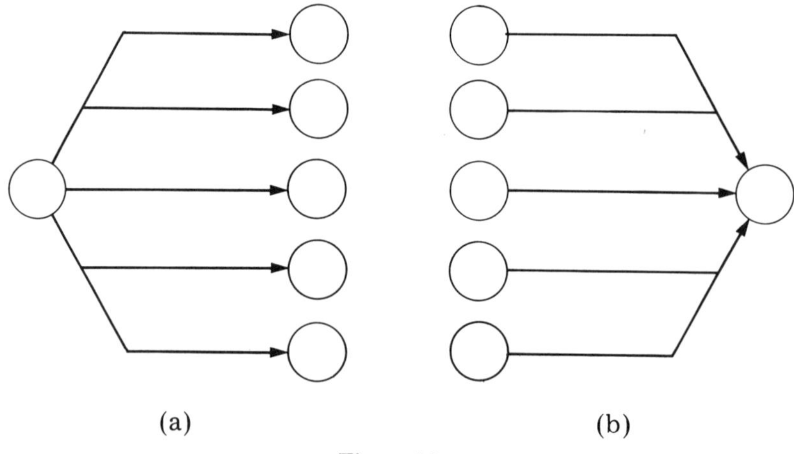

Figure 16.

Multiple Start and/or Finish Events

In most cases when network planning techniques are used on a project, a network diagram is used for all phases of the project from the very beginning until the project objectives are accomplished, and, therefore, these networks have one **Network Start** event and one **Network Finish** event. A network start event is an event that has no predecessor activities entering into it and has only activities leading from it. A network finish event is an event that has only activities entering into it and does not have activities leading from it.

Certain projects may have more than one network start event, or network finish event, or both. For example, a project may have multiple network finish events if it has several independent project objectives, rather than just one project objective. A project may have a research phase and a production phase. If a network diagram was developed for only the production phase, the diagram would show a multiple start network because there may be several interface or transition points between the research phase and production phase. For example, there may be several independent pieces of research information that are needed at different points in time in order for different production activities to start.

QUESTION 3. Draw a network plan that has one network start event that is the start event for Activities A, B, and C. When activity A is finished, activity D can start; when B is finished, F can start; when B and D are completed, E can start. Finally, the network has one network finish event which is the ending event of activities C, E, and F.

Drawing the Network Diagram

Having the project objectives, a list of activities, and a knowledge of network principles, you can begin drawing a network diagram. Start with the network start event and draw the activities in their logical precedence relationships as the project would progress toward the project objective. When deciding on the sequence in which the activities should be drawn to show their logical precedence relationship to one another, the following three questions should be asked regarding each individual activity:

1. Which activities must be completed *immediately* before this activity can be started?
2. Which activities can be done concurrently with this activity?
3. Which activities cannot be *immediately* started until this activity is completed?

By answering these questions for each activity, one should be able to draw a network diagram that visually shows the interrelationships of activities that are required to accomplish the project objective.

While drawing the network diagram, dummy activities should be added whenever necessary to help show the correct logical relationships. Also, after the network diagram is drawn, it should be checked to eliminate any loops.

The entire network diagram should flow from left to right, although some arrows for some individual activities may flow from right to left to prevent the overall network diagram from becoming too large. It should also be noted that, unlike the bar chart, the network diagram is *not* drawn to a time scale.

One may prefer to draw what is known as a **summary** network first, and then expand this summary network into a more detailed network. A summary network is one that contains a small number of generalized or high-level activities rather than a large number of detailed activities. A person may find it easier to first develop a list of general or high-level activities and draw a summary network, rather than to start immediately with developing a list of detailed activities and drawing the detailed network. In some cases, a summary network will suffice for use throughout the project, while in other cases a very detailed network may be desired for use during the project.

It is important to decide what level of detail is necessary for a particular project. Whatever the level of detail used in the initial network diagram, the user may still decide to further break up some activities into more detail as the project progresses. The level of detail may be based on the fact that there may be certain obvious breaking points between activities; for example, where there is a change in responsibility from one activity to the next activity, or the completion of a tangible "output" such as a report, drawing, or shipment of an item. In our example of the outhouse, the activity "buy materials" may be broken down into several more detailed activities, such as buy lumber or buy paint.

In some cases, a particular organization may perform similar projects time after time, and these projects may have certain portions that have the same type of activities in the same type of logical precedence relationship. In such cases, it may be worthwhile to have standard subnetworks for these portions of the projects. By having standard subnetworks, some effort and time can be saved when developing a network diagram for the overall project. Standard subnetworks should be drawn only for functions in which the logical relationships among the activities have been well established through historical practice, but these subnetworks should be flexible enough to be modified for a particular or unique project if the situation arises.

It is preferable to draw the network diagram on one large sheet of paper. Very large projects may require several or many sheets of paper. When more than one sheet of paper is used, the user must be careful to make proper cross-references for activities whose predecessor and successor events are on different pages. It may be a good idea to have standard subnetworks on separate pages.

QUESTION 4. What is a summary network?

After the network diagram has been drawn, it is necessary to number all the events in the network so that each activity in the network can be identified by its predecessor–successor event number combination. For example, in Figure 6 on page 11 activity A can be identified as 1-3, activity B as 2-3, and activity C as 3-4. No two events can have the same number. Each event must have its own unique event number.

Do not be too concerned about drawing the network diagrams neatly the first time. It is better initially to sketch out a rough draft of the network and make sure that the logical relationships of the activities are correct, and then go back and draw a neater looking diagram.

Referring to the example of building an outhouse in Appendix A, it is assumed that resources are available to perform simultaneously all tasks that are logically possible. The logic of the example starts as follows:

As the project starts, three activities can take place somewhat independently: deciding on the style (1-3), collecting 25 pounds of old newspaper per person (1-16), and testing the wind direction and choosing a location (1-2). Once a style is chosen, the plans can be drawn (3-4). Once a location has been chosen, it is possible to clear the land (2-4). When the land has been cleared and the plans drawn, digging the hole (4-12), buying the materials (4-5), and clearing a path from the main house (4-13) can all start.

QUESTION 5. Referring to the network diagram in Appendix A:

a. When the hole has been dug, the seat sanded, and the door mounted, which activity can then take place?

b. Which activities can be done in parallel with painting the exterior?

c. Which activity must be completed immediately prior to putting on the roof?

d. What is the ending event number for the activity "build frame"?

e. What is the number of the network finish event?

The network in Appendix A is drawn in what is known as the "activity-on-the-arrow" (AOA) format. There is another method of portraying activities known as the "activity-in-the-box" (or "activity-on-the-node," AON) or precedence diagramming method. Appendix B shows the outhouse example using the activity-in-the-box format. Using one method versus the other is a matter of personal preference. Both methods use a "network" and are based on precedence relationships among activities. The activity-in-the-box format has the advantage of not needing dummy activities and not using events.

Conclusion

The planning process involves defining a project objective, preparing a list of activities necessary to accomplish that objective, and developing a network diagram showing the activities in their logical precedence relationship. One of the most important benefits of network planning is that the preparation of the network diagram forces a person to plan on a systematic and logical basis. Another benefit resulting from drawing a network diagram is that alternate courses of action that were not previously obvious may come into focus as the network is being developed.

It should be mentioned that often there is no one correct or "best" network diagram for a particular project. For example, if five different people were given a project objective and asked to develop a network diagram for that project, most likely all five networks would vary to some degree. The network diagram is only a tool used to give a visual or graphical representation of one's *own* ideas of how to meet the project objective. All that can be said is that the only correct or "best" network diagram for a particular project is one that most realistically portrays the sequence in which the activities will actually be accomplished as the project progresses toward its objective.

Questions

1. What are the three basic parts of the planning function?
2. Define:
   a. Activity
   b. Event
   c. Loop
   d. Dummy
   e. Network start event
   f. Network finish event
3. For each statement in column A, choose the correct answer from column B.

A

(1) ______ start of an activity

(2) ______ not allowed

(3) ______ dashed arrow

(4) ______ should be specific and measurable

(5) ______

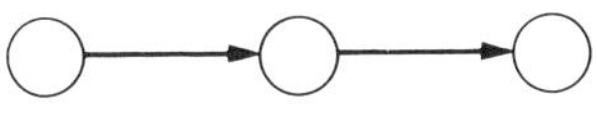

(6) ______ no two can be the same

(7) ______

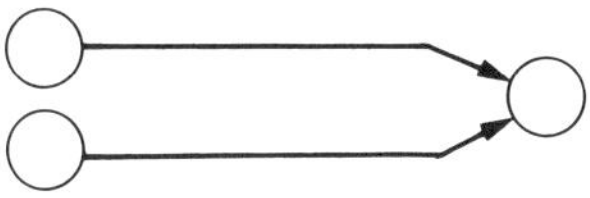

B

a. Dummy activity
b. Successor event
c. Parallel relationship
d. Network start event
e. Baselines
f. Series relationship
g. Predecessor event
h. Loops
i. Summary network
j. Network finish event
k. Project objectives
l. Laddering
m. Event number

4. Draw the network represented by the following information: The network start event is the beginning event for activities A and B. When A is finished, then both C and D can start. When

B is finished, then both E and F can start. When C and E are finished, then G can start. The project is complete when activities D, F, and G are finished.

5. Draw the network, and number the events to represent the following list of activities:

ACTIVITY	START EVENT	FINISH EVENT
A	1	2
B	1	3
C	1	5
D	2	4
E	2	6
F	3	4
dummy	3	5
G	3	6
H	4	6
I	5	6

6. How many things are wrong with the following network diagram?

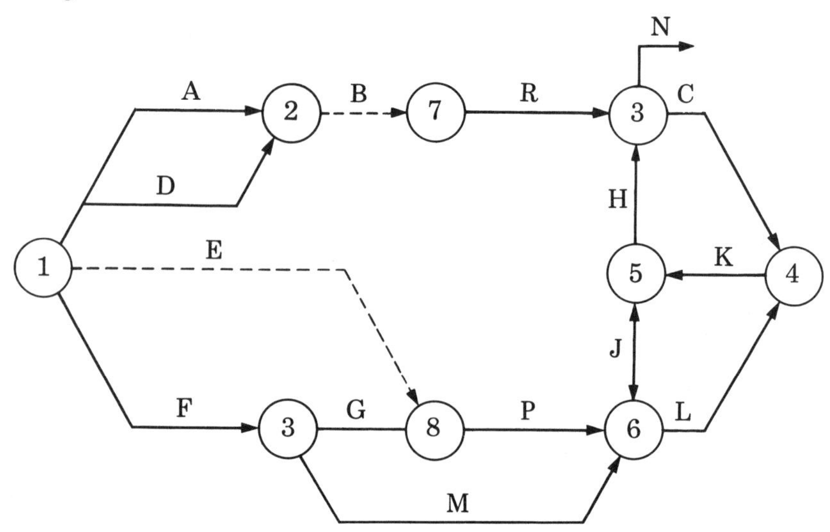

CHAPTER 3

Scheduling

The functions of planning and scheduling can be considered separately when using network planning techniques, but the scheduling function does depend on the planning function, since a project cannot be scheduled until it is first planned. The scheduling process begins with making an estimate of the duration of each activity in the network. Next, two scheduled times, the expected starting time of the project and the desired completion time of the project, are chosen. Based on activity duration estimates and scheduled times for the network start and finish events, calculations are made to find: (1) the *earliest* time that each activity is *expected* to start and finish, and (2) the *latest* time that each activity is *allowed* to start and finish.

The difference between the earliest expected times and the latest allowable times indicates the status of the overall project and its individual activities.

Activity Duration Estimates

The scheduling process can begin when the network diagram that resulted from the planning function has been prepared. The initial task of the scheduling function is to make estimates of the duration of each activity. These estimates do not refer to applied time, but rather to *total activity duration* or total elapsed time (applied time plus any nonapplied or waiting time).

It is good practice to have the person who will be responsible for accomplishing a particular activity make the duration estimate for that activity. This generates a commitment from the responsible person and prevents any bias that may occur from using the same estimator for all the activities.

When estimating durations, activities should be selected at random from the network rather than sequentially along any one path or chain of serial activities. This practice will help overcome any bias that may occur due to the fact that the estimator may be mentally keeping track of the cumulative duration that might be consumed by a chain of serial activities.

A consistent time base, such as hours, days, or weeks, should be used for the duration estimates for all the activities. When making estimates of durations, reference to calendar dates should be avoided.

For projects whose estimated activity durations are uncertain, it is possible to use three estimates for each activity:

1. Optimistic Time (t_o)—The time in which a particular activity may be completed if everything goes well and there are no complications. A rule of thumb is that there should be only one chance in ten of accomplishing the activity in less time than the optimistic time estimate.

2. Most Likely Time (t_m)—The time in which a particular activity can most often be completed under normal conditions. If an activity is repeated many times, the duration of time to accomplish this activity that occurs most frequently or most often would be equivalent to the most likely time estimate.

3. Pessimistic Time (t_p)—The time in which a particular activity may be completed under an adverse situation, such as having unusual and unforeseen complications. A rule of thumb is that there should be only one chance in ten of accomplishing the activity in more time than the pessimistic time estimate.

By using three time estimates, it is possible to take "uncertainty" into account when estimating how long an activity will take. The most likely time estimate must be longer than or equal to the optimistic time estimate, and the pessimistic time estimate must be longer than or equal to the most likely time estimate.

QUESTION 6. Match the following duration estimates for a particular activity:

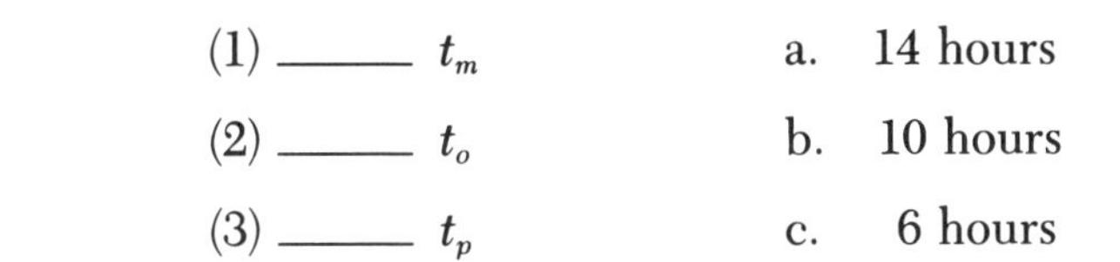

(1) ______	t_m	a.	14 hours
(2) ______	t_o	b.	10 hours
(3) ______	t_p	c.	6 hours

The three duration estimates for each activity should be marked on the network diagram in the general format shown in Figure 17a.

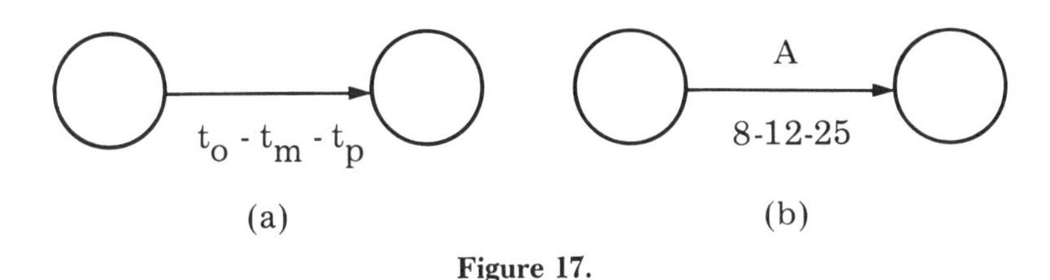

Figure 17.

Figure 17b shows that activity A has optimistic, most likely, and pessimistic estimates of 8, 12, and 25 days (or whatever time base is being used), respectively.

Looking at the outhouse network diagram in Appendix C (the reader should paper clip Appendix C, since reference will be made to it throughout this chapter and the next), you can see the three duration estimates for each activity. For eample, the activity 4-12, digging the hole, has an optimistic time estimate of 5 days, a most likely time estimate of 10 days, and a pessimistic time estimate of 30 days. In this case the pessimistic estimate of 30 days takes into account the possibility of running into large rock when digging the hole. Similarly, activity 13-16 has a pessimistic time estimate of 15 days, since it may take that long to paint the exterior if there are unfavorable weather conditions.

It should be noted that it is not necessary to give a time estimate for dummy activities, since, by definition, their duration is zero.

It is not required to give three time estimates for each activity. If one has good experience or data on how long it took to perform similar activities on past or completed projects, it may then be preferable to only make one time estimate for how long an activity is expected to take. Using the three-estimate approach (t_o, t_m, t_p) can be helpful when there is a high degree of uncertainty for how long an activity may take.

QUESTION 7. Referring to Appendix C, what are the optimistic, most likely, and pessimistic estimated durations for activity 6-7, "put on roof"? For activity 9-11, "dummy"?

QUESTION 8. What is the purpose of using three estimates for the duration of activities rather than just one estimate?

Event Scheduled Times (ST)

Having made three estimates of duration for each activity in the network, the next step is to choose a scheduled time for when the project is expected to start, and also a scheduled time for when the project is required to be completed. It should be noted that these are only *scheduled times*, and are *not necessarily the actual times* that the project will be started and completed. Scheduled times are specific points in time. They refer to the specific point in time when an event is scheduled to occur. Scheduled times do *not* refer to the duration of an activity; the latter is a period of time—a duration rather than a point in time. A scheduled time must be assigned to the network start event and network finish event. If the network has more than one network start event and/or network finish event, then a scheduled time must be assigned for each of these events.

A **Milestone Event** is one which, for some reason, is considered to be an important or key event in the network. For example, a certain event may show the start of a subnetwork. A network may have none, one, or several milestone events. The person planning the network must decide if any events in the network are to be milestone events. If a network has milestone events, a scheduled time may be assigned to these events. A scheduled time for a milestone event is the time that you want that event to occur.

For the outhouse example, event 1, the network start event, has been assigned a scheduled time of June 15, and event 13 is considered to be a milestone event, since it is the start of the painting, activities 13-14 and 13-16. Event 13 has been assigned a scheduled time of August 14.

QUESTION 9. Referring to the outhouse example in Appendix C, what is the scheduled time for the network finish event?

Schedule Calculations

Having the activity duration estimates and the event scheduled times, it is now possible to calculate a schedule for the entire project plan. The first step in performing the calculations involves obtaining an expected (or average) duration for each activity from its three time estimates. Next, calculations are made to find the earliest expected start and finish times for each activity. Finally, calculations are performed to obtain the latest allowable start and finish times for each activity.

Activity Expected Duration: t_e

When using three estimates of duration for each activity, it is a basic assumption of network planning that the three estimates for each activity follow a *Beta probability distribution*. With this assumption, it is possible to calculate an expected (also called mean or average) duration for each activity from the activity's three time estimates. This is accomplished by using the following formula:

$$t_e = \frac{t_o + 4(t_m) + t_p}{6} \tag{1}$$

Looking at our outhouse network example, the Beta probability distribution for activities 1-3, deciding style, and 1-2, testing wind and choosing location, are shown in Figures 18a and b, respectively.

By using Equation (1), the expected duration for activity 1-3 in Figure 18a is

$$t_e = \frac{1 + 4(5) + 15}{6} = 6 \text{ days}$$

For activity 1-2 in Figure 18b, the expected duration for activity is

$$t_e = \frac{10 + 4(15) + 20}{6} = 15 \text{ days}$$

which coincidentally happens to be the same as the most likely time estimate.

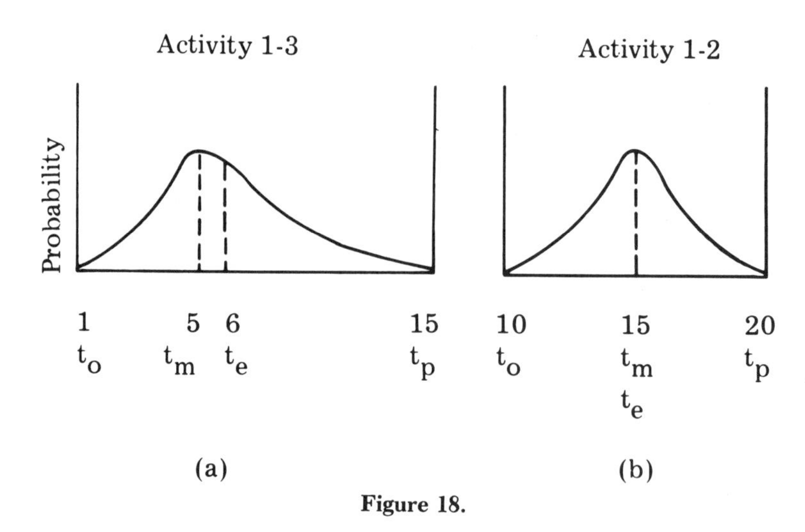

Figure 18.

The peaks of the curves in Figure 18 represent the most likely time estimate. The expected duration t_e divides the total area under the Beta probability curve into two equal parts. In other words, given any Beta probability curve, 50% of the area under this curve would be to the left of t_e and 50% would be to the right. For example, Figure 18a shows that 50% of the area under the curve is to the left of six days and 50% of the area is to the right of six days. Thus, it can be stated that there is a 50–50 chance that an activity actually will take more or less time than its expected duration. Or, there is a probability of 0.5 that it will actually take longer than t_e, and a probability of 0.5 that it will actually take less time than t_e. Thus, Figure 18a shows that there is a 50% chance that deciding on the outhouse style will actually take longer than six days, and a 50% chance that it will actually take less time than six days.

QUESTION 10. Using Equation (1) on page 27, calculate the expected duration for an activity having

$$t_o = 8, \ t_m = 12, \text{ and } t_p = 22$$

It is assumed that as the project progresses some of the activities actually will take as long or shorter than their expected duration, and other activities will take as long or longer than their expected duration. It is assumed further that by the time the entire project is completed the total net differences between all the *expected* durations and *actual* durations will be minimal.

Network planning can be considered a stochastic or probabilistic technique, since it allows for uncertainty in the activity duration by allowing three estimates which are assumed to be distributed according to the Beta probability distribution. On the other hand, the bar chart technique is considered a deterministic technique, since it allows only one time estimate per task.

QUESTION 11. a. What is the expected duration for the following activity?

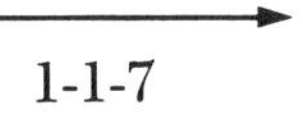

1-1-7

b. What are the chances of the activity actually taking less time than the expected duration? Of actually taking longer than expected?

In the outhouse network shown in Appendix C, the expected duration for each activity is shown in the box beneath each of the activities. It should be noted that, to make the calculations that will follow a little bit easier, the expected times were rounded off to the nearest whole day, rather than carrying the fraction.

Earliest Expected Start and Finish Times: ES and EF

Having an expected duration t_e for each activity in the network, and using the scheduled time for the network start event as a reference point, it is possible to calculate two points in time for each activity:

1. *Earliest (Expected) Start Time (ES)*—The earliest time by which a particular activity is expected to begin, based on calculations made from the scheduled time of the network start event and from the expected durations of preceding activities.

2. *Earliest (Expected) Finish Time (EF)*—The earliest time by which a particular activity is expected to be completed. It is equal to the activity's earliest expected start time plus the activity's expected duration. ($EF = ES + t_e$.)

The earliest times, ES and EF, are found by calculating *forward* through the network, that is, by working from the network start event toward the network finish event. As these ES and EF calculations are performed, the following rules should be kept in mind:

Rule 1. The earliest start time for an activity or activities leading from a particular event must be equal to or later than the latest of all the earliest finish times of all the activities entering into that same event.

Rule 2. The earliest finish time for a particular activity is equal to its earliest start time plus its expected duration.

$$EF = ES + t_e \tag{2}$$

The outhouse example in Appendix C will be used to illustrate the ES and EF calculations. The calculations will asssume a five-day work week, Monday through Friday, and therefore no Saturdays and Sundays will show up as an earliest start or finish time. Also, no work will be performed on Labor Day, September 7. (Note that all dates calculated will be based on the calendar on page 31.)

Activity 1-2, testing wind and choosing location, has as its beginning event, the network start event with a scheduled date of June 15. Therefore, the earliest start date for activity 1-2 is also June 15. Applying rule 2, above, the earliest finish date for this activity will be July 3, which is 15 working days after its earliest start date.

Activity 1-3 will also have an ES of June 15 and will have an EF on June 22, which is six working days beyond June 15.

Collecting newspapers, activity 1-16, will also have an ES of June 15, and its earliest finish date will be June 26.

Clearing land, activity 2-4, cannot begin until activity 1-2 is completed. Since activity 1-2 has an earliest finish date of July 3, then activity 2-4 will have an earliest start date of July 6 (July 4 and July 5 are a weekend). The EF for activity 2-4 will be 11 work days beyond July 6, or July 20.

Similarly, activity 1-3 has an EF of June 22, and therefore activity 3-4 cannot be expected to start until June 23. With an ES of June 23, the EF for drawing the plans will be June 25.

Rule 1, above, is applicable to activities 4-5, 4-12, and 4-13. These three activities cannot start until both activities 2-4 *and* 3-4 are completed. Activity 2-4 has an EF of July 20, and activity 3-4 has an EF of June 25. Thus, event 4 is not expected to occur until July 20, which is the later of these two earliest finish dates. Therefore, activities 4-5, 4-12, and 4-13 will all have an ES of July 21.

	S	M	T	W	T	F	S
		1	2	3	4	5	6
	7	8	9	10	11	12	13
JUNE	14	15	16	17	18	19	20
	21	22	23	24	25	26	27
	28	29	30				
				1	2	3	4
	5	6	7	8	9	10	11
JULY	12	13	14	15	16	17	18
	19	20	21	22	23	24	25
	26	27	28	29	30	31	
							1
	2	3	4	5	6	7	8
	9	10	11	12	13	14	15
AUGUST	16	17	18	19	20	21	22
	23	24	25	26	27	28	29
	30	31					
			1	2	3	4	5
	6	7	8	9	10	11	12
SEPTEMBER	13	14	15	16	17	18	19
	20	21	22	23	24	25	26
	27	28	29	30			

QUESTION 12. What will be the earliest finish date for activity 4-5? Activity 4-12? Activity 4-13?

QUESTION 13. Using the calendar on page 31, if the ES for a particular activity is July 6 and its expected duration is 12 work days, what is its EF, assuming a 6-day work week (exclude Sundays)? Assuming a 7-day work week?

The earliest start and finish dates for the remaining activities in the network are shown in Figure 19 on page 35.

Latest Allowable Finish and Start Times: LF and LS

Having the expected duration t_e for each activity in the network and using the scheduled time for the network finish event as a reference point, it is possible to calculate for each activity its

1. *Latest (Allowable) Finish Time (LF)*—The latest time by which a particular activity must be completed in order to ensure the completion of the project by the scheduled time of the network finish event. This time is based on calculations made from the scheduled time of the network finish event and from the expected durations of succeeding activities.

2. *Latest (Allowable) Start Time (LS)*—The latest time by which a particular activity must begin so that the project can be completed by the scheduled time of the network finish event. It is equal to the activity's latest allowable finish time minus the activity's expected duration. ($LS = LF - t_e$.)

The latest finish and start times are found by calculating *backward* through the network; that is, by working from the network finish event back to the network start event. As these LF and LS calculations are performed, the following rules should be kept in mind:

Rule 3. The latest finish time for an activity or activities entering into a particular event must be equal to or earlier than the earliest of all the latest start times of all the activities leading from that same event.

Rule 4. The latest start time for a particular activity is equal to its latest finish time minus its expected duration:

$$LS = LF - t_e \tag{3}$$

To illustrate the LF and LS calculations, the outhouse example in Appendix C will be used again. The calculations will assume a five-day work week, Monday through Friday, and therefore no Saturdays and Sundays will show up as a latest finish or start time. (Again, all dates calculated will be based on the calendar on page 31.)

Testing the outhouse, activity 16-17, has as its ending event, the network finish event with a scheduled date of August 31. Therefore, the latest finish date for activity 16-17 is also August 31. Applying rule 4, above, the latest start date for this activity will be August 19, which is nine working days before its latest finish date.

Collecting newspapers, activity 1-16, must be completed, along with activities 13-16, 14-16, and 15-16, before activity 16-17 can start. Thus, these four activities must all be completed by at least August 18, in order that activity 16-17 can start by August 19, and the entire project be completed by August 31. Therefore, activity 1-16 will have an LF of August 18 and an LS of August 5, since its expected duration is ten days.

Painting the exterior, activity 13-16, will also have an LF of August 18, and its LS will be August 10. The LS of activity 14-16 will be August 12. Activity 15-16 is a dummy activity and will have an LF of August 18. Because a dummy activity has a zero duration, its LS will also be August 18.

Since activity 14-15, installing the door latch, is succeeded by a dummy activity in which no work is expended, it will have a latest finish date of August 18, which is the same as the LS for the dummy activity 15-16. Its LS, then, will be August 17.

Rule 3 is applicable to activity 13-14, painting the interior, which must be completed before activities 14-15 and 14-16 can start. Activity 14-15 has an LS of August 17, and activity 14-16 has an LS of August 12. Since the earlier of these two dates is August 12, activity 13-14 must be finished before this date, and so its LF will be August 11, and its LS will be August 6, which is four working days before its LF.

Looking at activities 4-13 and 12-13, rule 3 is again applicable.

These two activities must be completed before activities 13-14 and 13-16 can start.

The latest allowable finish and start dates for the remaining activities in the network are shown in Figure 19 on page 35.

QUESTION 14. What are the latest start and finish times for activities 4-13 and 12-13?

QUESTION 15. Use the calendar on page 31.

a. If the LF for an activity is August 8 and the expected duration is 17 work days, what is its LS, assuming a six-day work week (exclude Sundays)?

b. If the LS for an activity is September 15 and its expected duration is eight days, what would its LF be, assuming a five-day work week (no Saturdays and Sundays)?

Event Occurrence Times: ET and LT

Up to this point, ES, EF, LS, and LF times have been mentioned in reference to the activities in the network. From our previous definition of an event, we know that an event is said to occur when all the activities leading into that event have been completed; an event does not consume time, but rather it is an instant or point in time. Since an event is a point in time, we cannot talk about an earliest expected start time and earliest expected finish time for an event because they are one and the same, so, when referring to an event, we can only talk about an *Earliest (Expected) Time* (ET) to occur. In like manner, we cannot talk about a latest allowable start time and latest allowable finish time for an event because they, too, are one and the same. We can only talk about a *Latest (Allowable) Time* (LT) to occur.

An event's earliest time (ET) to occur is equal to the latest time of all the earliest finish times of all the activities entering into that event. For example, event 4 has two activities, 2-4 and 3-4, entering into it; the earliest finish dates of these two activities are July 20 and

PRED. EVENT	SUCC. EVENT	ACTIVITY DESCRIPTION	t_e	ES	EF	LS	LF	TOTAL SLACK	FREE SLACK
1	2	Choose Location	15	6/15	7/3	6/2	6/22	−9	0
1	3	Decide on Style	6	6/15	6/22	6/25	7/2	8	0
1	16	Collect Newspaper	10	6/15	6/26	8/5	8/18	37	46
2	4	Clear Land	11	7/6	7/20	6/23	7/7	−9	0
3	4	Draw Plans	3	6/23	6/25	7/3	7/7	8	17
4	5	Buy Materials	5	7/21	7/27	7/8	7/14	−9	0
4	12	Dig Hole	13	7/21	8/6	7/16	8/3	−3	6
4	13	Clear Path	7	7/21	7/29	7/28	8/5	5	14
5	6	Build Frame	4	7/28	7/31	7/15	7/20	−9	0
5	10	Build Door	1	7/28	7/28	7/29	7/29	1	0
6	7	Put On Roof	1	8/3	8/3	7/21	7/21	−9	0
7	8	Put On Exterior Siding	2	8/4	8/5	7/22	7/23	−9	0
8	9	Put Up Interior Walls	5	8/6	8/12	7/24	7/30	−9	0
9	11	Dummy Activity	0	8/12	8/12	7/30	7/30	−9	0
9	12	Sand Seat	1	8/13	8/13	8/3	8/3	−8	1
10	11	Cut Half Moon in Door	1	7/29	7/29	7/30	7/30	1	10
11	12	Mount Door	2	8/13	8/14	7/31	8/3	−9	0
12	13	Place Bldg. over Hole	2	8/17	8/18	8/4	8/5	−9	0
13	14	Paint Interior	4	8/19	8/24	8/6	8/11	−9	0
13	16	Paint Exterior	7	8/19	8/27	8/10	8/18	−7	2
14	15	Install Door Latch	2	8/25	8/26	8/17	8/18	−6	0
14	16	Hang Pictures	5	8/25	8/31	8/12	8/18	−9	0
15	16	Dummy Activity	0	8/26	8/26	8/18	8/18	−6	3
16	17	Test	9	9/1	9/14	8/19	8/31	−9	0

Figure 19. Activity Schedule.

June 25, respectively. The later of these two dates is July 20, so the earliest time that we can expect event 4 to occur is July 20.

An event's latest time (LT) to occur is equal to the latest finish time of all the activities entering into that event. Again looking at event 4, the LF for activities 2-4 and 3-4 is July 7. Therefore, the latest time that this event must occur is July 7. If event 4 does not occur by its LT, then we jeopardize completing the project by August 31, which is the scheduled date for the network finish event.

It should be noted that for a network start event, the earliest time to occur will be the same as its scheduled time; for a network finish event, the latest time to occur will be the same as its scheduled time.

Figure 20 on page 37 shows the earliest dates and latest dates for each of the events in our example network. It is not necessary, however, to calculate an event schedule as in Figure 20 if only activity information is desired, such as start and finish times as shown in the activity schedule in Figure 19 on page 35.

QUESTION 16. Using the information in the activity schedule in Figure 19 on page 35, what are the earliest and latest occurrence dates for the following events?

EVENT	ET	LT
1		
5		
7		
12		
16		
17		

Slack

Slack (sometimes called float) is an indication of the status of a project and its individual activities. It is a measure of how far the project is ahead or behind the scheduled time of the network finish event. There are several different types of slack. **Total slack** (TS) and

free slack (FS) refer to activities, whereas **event slack** refers to the status of the events in the network.

EVENT	ET	LT	EVENT SLACK	ST
1	6/15	6/2	−9	6/15
2	7/3	6/22	−9	
3	6/22	7/2	8	
4	7/20	7/7	−9	
5	7/27	7/14	−9	
6	7/31	7/20	−9	
7	8/3	7/21	−9	
8	8/5	7/23	−9	
9	8/12	7/30	−9	
10	7/28	7/29	1	
11	8/12	7/30	−9	
12	8/14	8/3	−9	
13	8/18	8/5	−9	8/14
14	8/24	8/11	−9	
15	8/26	8/18	−6	
16	8/31	8/18	−9	
17	9/14	8/31	−9	8/31

Figure 20. Event Schedule.

Total Slack: TS

Total Slack is the maximum amount of time that the activities on a particular path (a chain of one or more serial activities) can be delayed without jeopardizing the completion of the project by the scheduled time of the network finish event. The total slack for a certain path in the network is *common* to all activities which are on that path, and therefore all those activities will have the *same* total slack. The total slack for a particular activity in the network is found by subtracting the activity's earliest finish time from its latest finish time, or it can be obtained by subtracting the activity's earliest start time from its latest start time. This is shown by the formula

$$\text{Total Slack} = \text{TS} = \text{LF} - \text{EF} = \text{LS} - \text{ES} \qquad (4)$$

Total slack has the same time units (weeks, days, or hours) as the activity's expected duration.

A large network will have many paths, each with a different amount of total slack, or in some cases more than one path will have identical total slack. The path with the least value (either least positive or most negative) of total slack is called the **Most Critical Path**. The most critical path can also be defined as the longest or most time-consuming path (chain of activities) from the network start event to the network finish event.

If the total slack of the most critical path is a positive value, it is an indication that the project has a better than 50–50 chance of being completed before the scheduled time of the network finish event. Also, if the most critical total slack is a positive value, then the earliest time for the network finish event to occur will be before or earlier than its scheduled or latest time to occur.

If the most critical total slack is zero, the earliest time for the network finish event to occur will be the same as its scheduled time. A total slack of the most critical path of zero indicates that the project has a 50% chance of being completed before the scheduled time of the network finish event, and a 50% chance of being completed after the scheduled time.

If the total slack of the most critical path is a negative value, it is an indication that the project has a less than 50–50 chance of being completed before the scheduled time of the network finish event. Also, if the most critical total slack is a negative value, the earliest time for the network finish event to occur will be after or later than its scheduled time.

Any path that has a negative value of total slack can be considered critical, and any path that has a positive value of total slack can be considered noncritical.

In referring to the network diagram in Appendix C and to its schedule in Figure 19 on page 35, it can be seen that the least value of total slack is -9 work days. The following activities *share* this most critical total slack: 1-2, 2-4, 4-5, 5-6, 6-7, 7-8, 8-9, 9-11, 11-12, 12-13, 13-14, 14-16, and 16-17. These activities form the path 1-2-4-5-6-7-8-9-11-12-13-14-16-17, which is the most critical path in the network, since it is the longest or most time-consuming path. The total time for this path is found by totaling the expected duration of all the activities on the path. This total time is $15+11+5+4+1+2+$

5+0+2+2+4+5+9 = 65 working days. This path consumes more time than any other path from the network start event to the network finish event. Since the value of the most critical total slack is negative, there is a less than 50–50 chance of completing the outhouse by August 31. Furthermore, the earliest time for completion of the project would be September 14, which is 9 work days later than August 31. It must be emphasized that each activity on the most critical path does *not* have −9 work days total slack *individually*, but rather the −9 work days are *shared by all* the activities on the most critical path.

The schedule in Figure 19 on page 35 shows that the activities that are not on the most critical path have a total slack with a more positive value than the most critical total slack. For example, activities 1-3 and 3-4 have eight work days of total slack. Each of these activities is expected to be completed (EF) eight work days ahead of or before they must be finished (LF). These two activities form the path 1-3-4, and the eight work days of total slack are shared by both of the activities on the path.

QUESTION 17. If the most critical slack in a network is a positive value, when can the project be expected to be completed?

QUESTION 18. If a network has an Earliest Occurrence date of October 15 and an LT of October 1 for its network finish event, would the chances of completing the project by its scheduled date be less than or greater than 50–50?

Free Slack: FS

Free Slack may be defined as the amount of time that a particular activity can be delayed without delaying the earliest start time of any *immediately* succeeding activities. It is the *relative difference* in the amount of total slack between the activity with the least value of total slack entering into a given event and the values of total slack of all the activities entering into that same event. Free slack is calculated by subtracting the least value of total slack of all the activities entering into a particular event from the values of total slack of all the activities entering that event. Since free slack is the relative difference between total slacks of activities entering an event, it will only exist when there are two or more activities entering a particular event. Also, since free slack is the relative difference from the least total slack value entering the event, it will *never* be a *negative* value.

Looking at the outhouse example, event 4 has two activities, 2-4 and 3-4, entering into it. The total slack values of these two activities are -9 and 8 work days, respectively. The lesser of these two values is -9, and thus the free slack for activity 3-4 is found by: $8-(-9) = 17$ work days. [It should be noted that activity 2-4 has a free slack of $-9-(-9) = 0$.] This value of 17 work days tells us that activity 3-4 can be delayed as much as 17 work days without causing a delay in the earliest start times of activities 4-5, 4-12, and 4-13. This is easy to see since these three activities cannot start until both 2-4 and 3-4 are completed, and 2-4 has an EF of July 20, while 3-4 has an EF of June 25, which is 17 work days before July 20. Therefore, activity 3-4 can be delayed 17 work days, or up to until July 20, without causing a delay in the earliest expected start of activities leading from event 4.

Event Slack

Unlike the two previously mentioned slacks, which referred to activities, **Event Slack** refers, as its name implies, to events. Event slack may be defined as the maximum amount of time that the occurrence of a particular event may be delayed without jeopardizing the completion of the project by the scheduled time of the network finish event. Event slack is calculated by subtracting an event's expected time to occur from its latest occurrence time:

$$\text{Event Slack} = LT - ET \qquad (5)$$

Event slack for a parituclar event is also equal to the least value of total slack of all the activities entering into that event. For example, event 12 has three activities, 4-12, 9-12, and 11-12, entering into it. These three activities have total slack values of -3, -8, and -9 work days, respectively. The least of these values is -9, and therefore the event slack for event 12 is equal to -9 work days. The event slack for the other events in our example is shown in Figure 20 on page 37.

QUESTION 19. Referring to the following diagram and assuming that the total slack values for activities A, B, C, D, and E are −12, −4, 0, +8 and +21, respectively, what is the free slack for each activity? What is the event slack for event number 99?

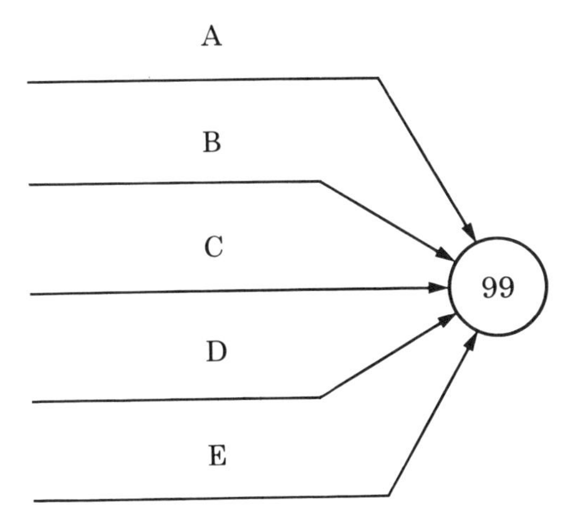

Conclusion

The scheduling function depends on a plan having been developed beforehand. The scheduling process involves developing a timetable for the plan, but in order to develop a schedule, estimates for the duration of each of the activities in the network must be made, and also scheduled times must be chosen for when the project is expected to start and when it must finish. Once these estimates are made, a schedule is calculated that shows the earliest expected start and finish times and the latest allowable start and finish times for each activity in the network. When these times have been calculated, it is possible to calculate various types of slack, which are used to indicate the status of the overall project as well as of the individual activities. Being able to report on the status of the project is one of the primary purposes of developing a schedule.

Appendix D shows the outhouse example, in the activity-in-the-box format, with the schedule input data for each activity. The number in the lower left-hand corner is the **activity number**, and the number in the lower right-hand corner of each box is its expected duration estimate. Figure 21 on page 43 shows the associated schedule. Note that since the activity-in-the-box format does not have events, the schedule indicates the **activity number** as well as the activity number of those activities that are immediate predecessors.

Questions

1. True or False?

   a. __________ The length of an activity arrow is determined by its expected duration.

   b. __________ The pessimistic time estimate is less than or equal to the optimistic time estimate.

   c. __________ With network planning, you must always use three time estimates.

   d. __________ Every network has a milestone event.

   e. __________ Sometimes an activity is actually accomplished in less time than its expected duration.

   f. __________ $EF = ES + t_e$

   g. __________ The latest start times are calculated working "forward" through the network.

   h. __________ $LS = LF + t_e$

   i. __________ ET and LT always refer to events.

   j. __________ Event slack is always positive.

2. Define:

   a. Optimistic time estimate.

   b. Earliest start time.

   c. Latest finish time.

   d. Earliest expected occurrence time.

   e. Total slack.

ACTIVITY NUMBER	PREDECESSOR ACTIVITIES	ACTIVITY DESCRIPTION	t_e	ES	EF	LS	LF	TOTAL SLACK	FREE SLACK
1		Choose Location	15	6/15	7/3	6/2	6/22	−9	0
2		Decide on Style	6	6/15	6/22	6/25	7/2	8	0
3		Collect Newspaper	10	6/15	6/26	8/5	8/18	37	46
4	1	Clear Land	11	7/6	7/20	6/23	7/7	−9	0
5	2	Draw Plans	3	6/23	6/25	7/3	7/7	8	17
6	4, 5	Buy Materials	5	7/21	7/27	7/8	7/14	−9	0
7	4, 5	Dig Hole	13	7/21	8/6	7/16	8/3	−3	6
8	4, 5	Clear Path	7	7/21	7/29	7/28	8/5	5	14
9	6	Build Frame	4	7/28	7/31	7/15	7/20	−9	0
10	6	Build Door	1	7/28	7/28	7/29	7/29	1	0
11	9	Put on Roof	1	8/3	8/3	7/21	7/21	−9	0
12	11	Put On Exterior Siding	2	8/4	8/5	7/22	7/23	−9	0
13	12	Put Up Interior Walls	5	8/6	8/12	7/24	7/30	−9	0
14	13	Sand Seat	1	8/13	8/13	8/3	8/3	−8	1
15	10	Cut Half Moon in Door	1	7/29	7/29	7/30	7/30	1	10
16	13, 15	Mount Door	2	8/13	8/14	7/31	8/3	−9	0
17	7, 14, 16	Place Bldg. over Hole	2	8/17	8/18	8/4	8/5	−9	0
18	8, 17	Paint Interior	4	8/19	8/24	8/6	8/11	−9	0
19	8, 17	Paint Exterior	7	8/19	8/27	8/10	8/18	−7	2
20	18	Install Door Latch	2	8/25	8/26	8/17	8/18	−6	0
21	18	Hang Pictures	5	8/25	8/31	8/12	8/18	−9	0
22	3, 19, 20, 21	Test	9	9/1	9/14	8/19	8/31	−9	0

Figure 21. Activity Schedule (Activity-in-the-Box Format).

3. For each symbol in column A, choose the correct answer from column B.

	A		B
(1) ______	EF	a.	Optimistic time estimate
(2) ______	t_p	b.	Earliest start time
(3) ______	ET	c.	Latest time to occur
(4) ______	LS	d.	Pessimistic time estimate
(5) ______	t_m	e.	Free slack
(6) ______	TS	f.	Earliest finish time
(7) ______	LT	g.	Scheduled time
(8) ______	t_e	h.	Most likely time estimate
(9) ______	FS	i.	Total slack
(10) ______	t_o	j.	Latest finish time
(11) ______	ES	k.	Expected duration
(12) ______	LF	l.	Latest start time
(13) ______	ST	m.	Earliest time to occur

4. Fill in the table for the network on page 45. Assume all duration estimates are in days. Use the calendar on page 31 as reference for your calculations. Also assume a Monday to Friday work week.

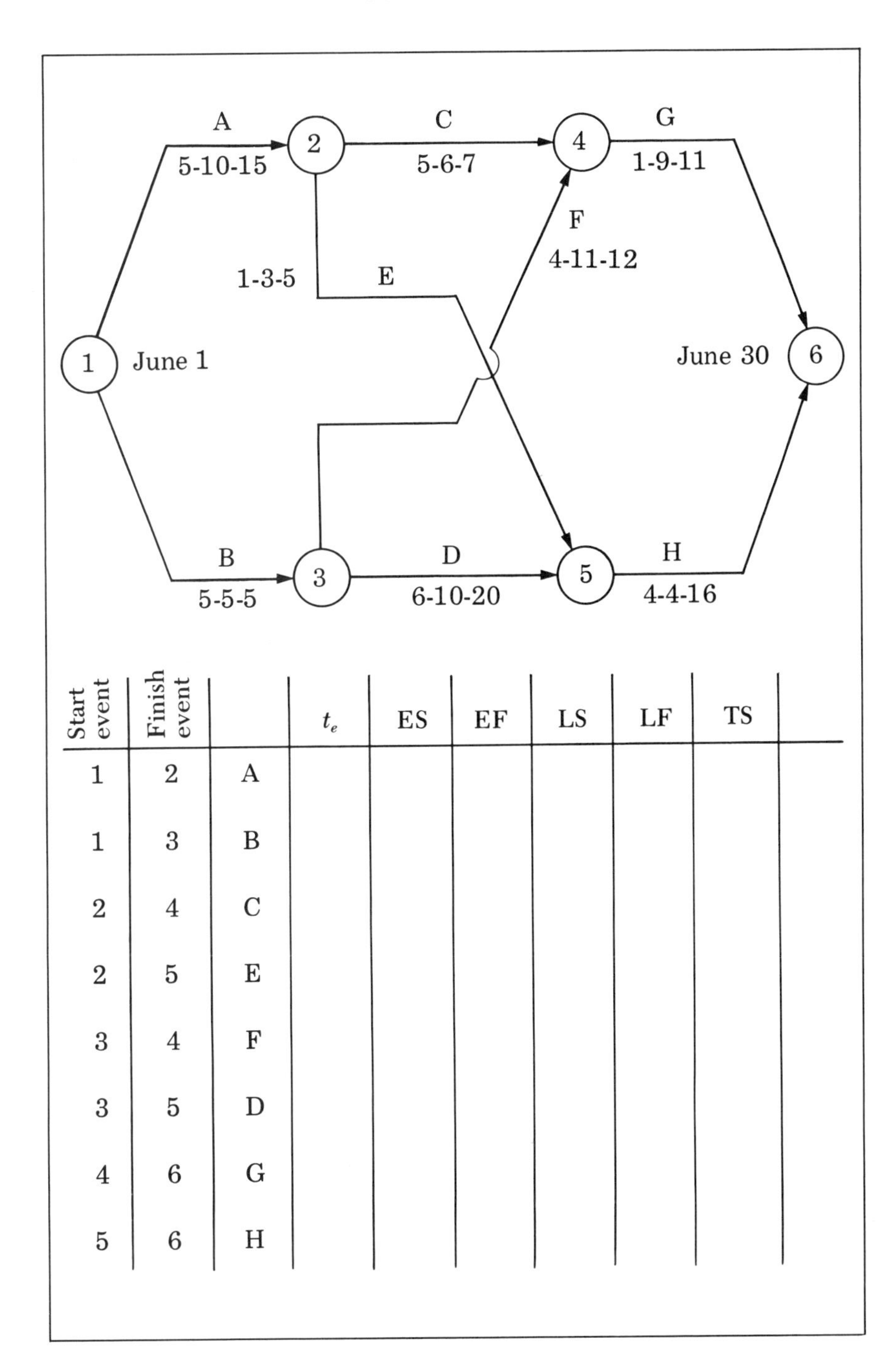

Start event	Finish event		t_e	ES	EF	LS	LF	TS
1	2	A						
1	3	B						
2	4	C						
2	5	E						
3	4	F						
3	5	D						
4	6	G						
5	6	H						

CHAPTER 4

Probability Considerations

(This chapter assumes that three time estimates have been used for the duration of each activity. If only one time estimate has been used for each activity, then probability calculations cannot be made.)

Since it is assumed that the three time estimates for each activity in the network follow a Beta probability distribution, it is possible to calculate the probability or likelihood of actually completing the project before the scheduled time of the network finish event.

Fundamentals

Having three time estimates (t_o, t_m, t_p) for a particular activity in the network, it is possible to obtain an expected duration, t_e, for that activity by using Equation (1) on page 27. There is a probability of 0.5 that the activity will actually take less time than its expected duration and a probability of 0.5 that it will actually take longer than t_e. Now, if all the activities on the most critical path are added together, a total probability distribution is obtained. From probability theory, the central limit theorem states that this total probability distribution is not a Beta probability distribution, but a *normal* probability distribution, which is bell-shaped and symmetrical about its expected value (also called the mean). Furthermore, this total probability distribution has an expected (duration) value that is equal to the *sum* of the expected durations of all the activities which make up the total distribution. It also has a variance that is equal to the sum of the variances of all the activities which make up the total distribution.

While the expected duration, t_e, for the Beta probability distribution of a particular activity's three time estimates ($t_o + t_m + t_p$) is obtained by using Equation (1) on page 27, the **variance** for the Beta probability distribution of the activity is found by the following formula:

$$\text{variance} = \sigma^2 = \left(\frac{t_p - t_o}{6}\right)^2 \tag{6}$$

Remember, the variance of the *normal* distribution is the *sum* of the variances of the Beta distribution.

QUESTION 20. Compute the expected duration and the variance for the following Beta probability distribution:

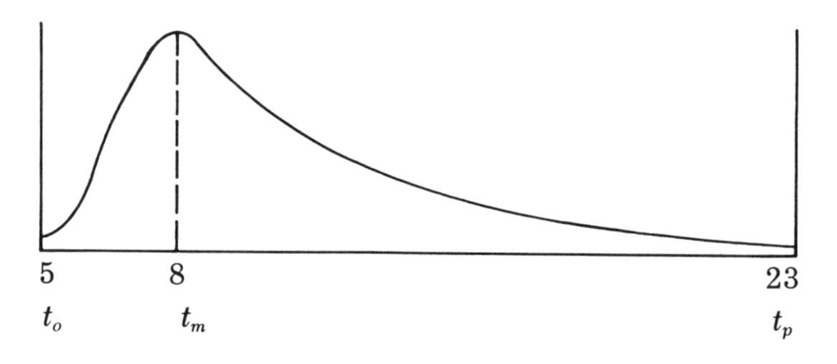

While the expected duration value, which divides the area under a probability distribution into two equal parts, is a measure of the central tendency of a distribution, the variance is a measure of the dispersion or spread of a distribution from its expected value. The **standard deviation**, σ, is another measure of the dispersion of a distribution and is *equal to the square root of the variance.* The standard deviation gives a better visual representation of the spread of a distribution from its expected value or mean than does the variance. For Figure 22, which is a normal distribution, one standard deviation to each side of the mean or expected value of a normal curve includes 68% of the total area under the normal curve, two standard deviations to each side of the mean include 95% of the area, and three standard deviations to each side of the mean include 99% of the total area under the normal curve.

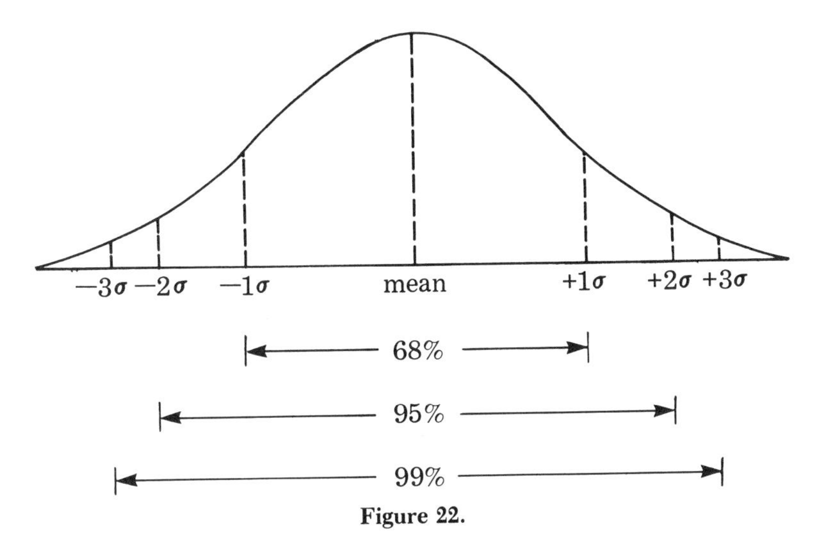

Figure 22.

The standard deviation is a measure of the dispersion or the amount of spread of a distribution. Figure 23 shows two normal distributions; Figure 23a is more widespread and thus has a larger standard deviation than does Figure 23b. However, in both Figures 23a and b, 68% of the area under each curve is included within one standard deviation of the respective means.

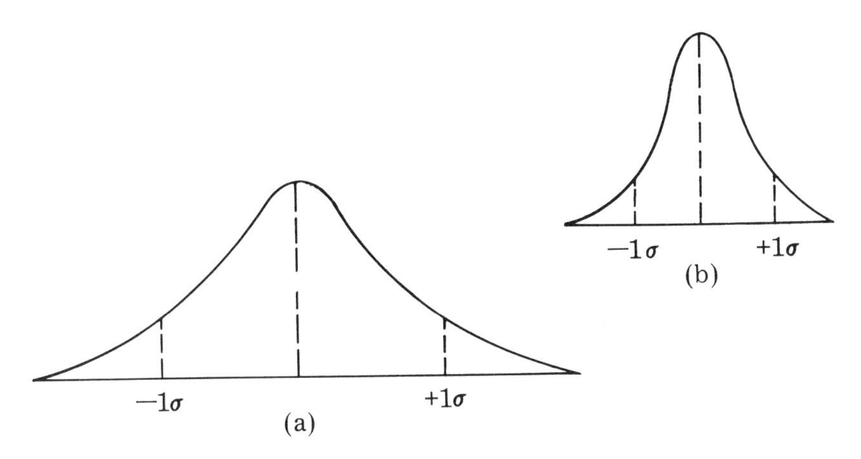

Figure 23.

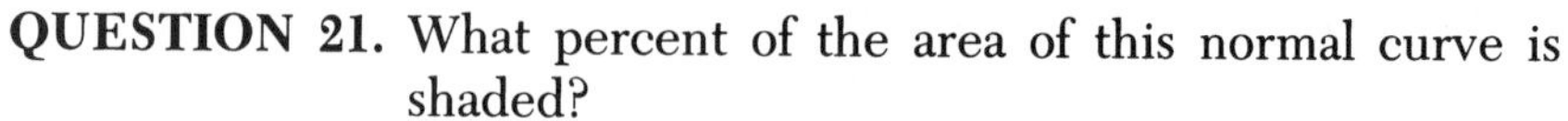

QUESTION 21. What percent of the area of this normal curve is shaded?

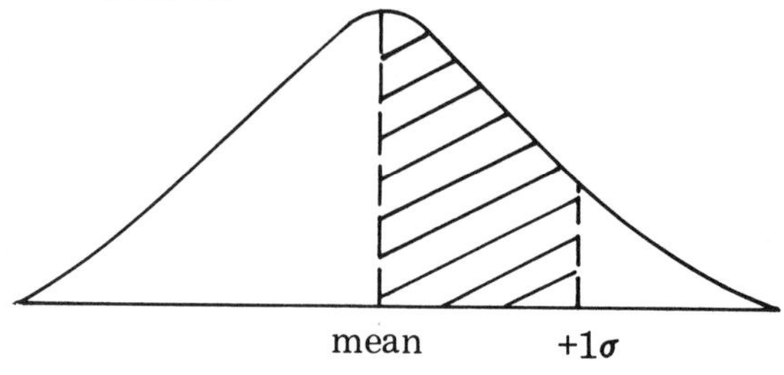

QUESTION 22. If 95% of the area under the following normal curve is between the two labeled points, what is the variance?

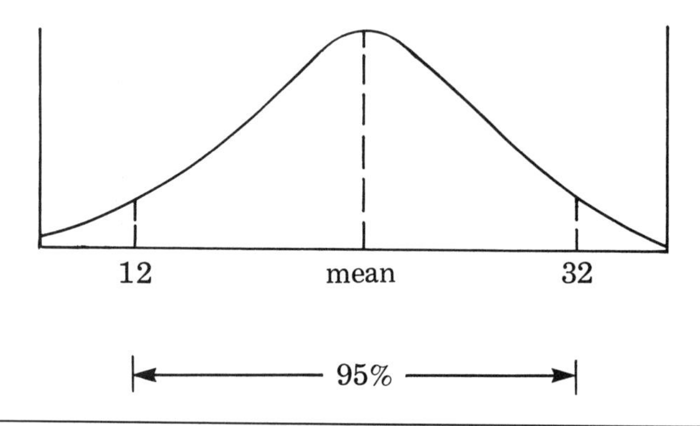

We now know that the total probability distribution of all the activities on the most critical path of a network is normally distributed with a mean equal to the sum of the individual activity durations and a variance equal to the sum of the individual activity variances. To illustrate this, consider the simple example network in Figure 24.

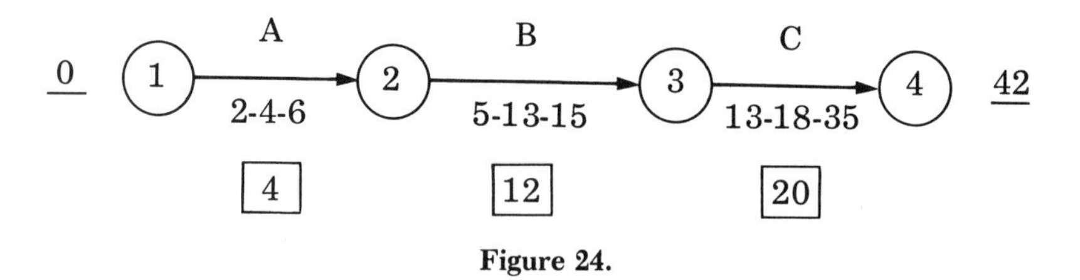

Figure 24.

The probability distributions for each of the activities in Figure 24 are shown in Figures 25a, b, and c.

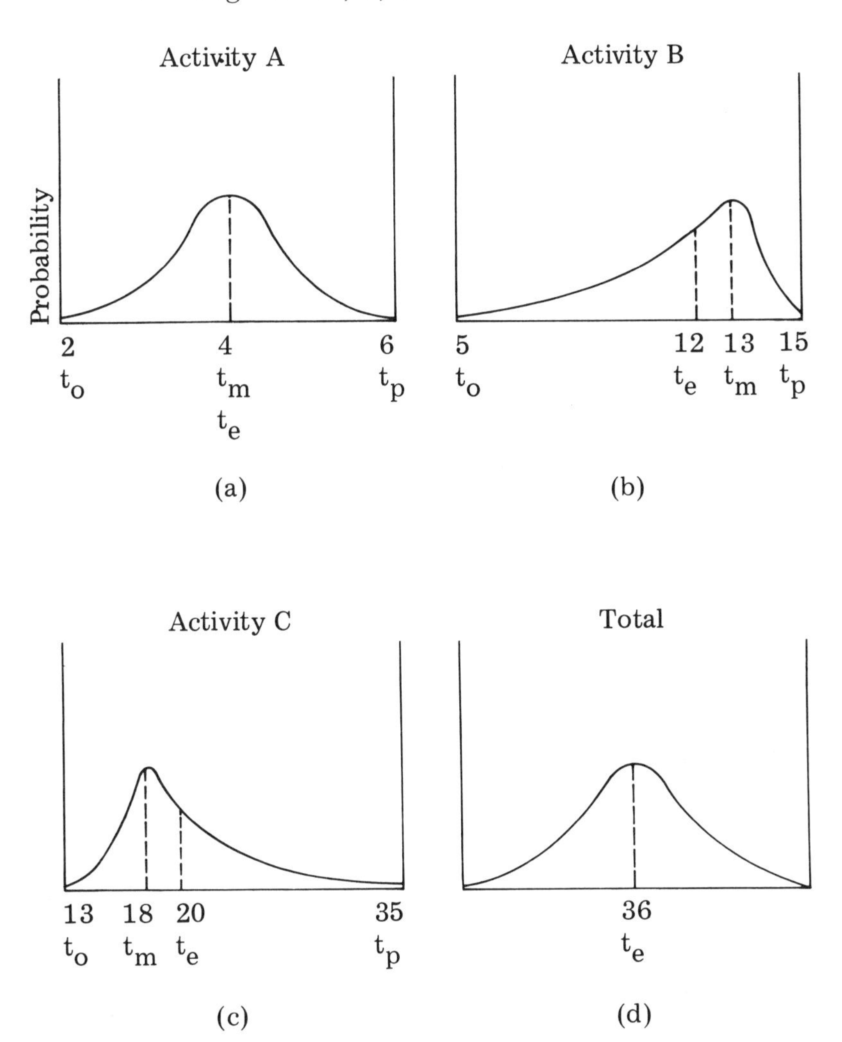

Figure 25.

The mean or expected duration for each activity is:

ACTIVITY		t_e
A	$\frac{2 + 4(4) + 6}{6}$	= 4 days
B	$\frac{5 + 4(13) + 15}{6}$	= 12 days
C	$\frac{13 + 4(18) + 35}{6}$	= 20 days
Total		36 days

Now if we sum the three distributions, we obtain a total mean or total t_e of:

ACTIVITY	t_o	t_m	t_p
A	2	4	6
B	5	13	15
C	13	18	35
Total	20	35	56

$$\text{Total } t_e = \frac{20 + 4(35) + 56}{6} = 36 \text{ days}$$

This is the same as the sum of the three individual t_e's previously calculated: 4 + 12 + 20 = 36 days. The total probability distribution is shown in Figure 25d on page 51. Thus, the total expected duration for the path 1-2-3-4 is 36 days. The network has a scheduled time of 0 for the network start event, and a scheduled time of 42 days for the network finish event. The earliest expected occurrence time for the network finish event will be 36 days beyond the start time of activity A.

The total distribution has a mean elapsed time equal to the sum of the three individual t_e's; there is a probability of 0.5 that the project will actually be completed before 36 days and a probability of 0.5 that it will actually be completed after 3 days. Applying the same principle to the outhouse example in Figure 20 on page 37, it can then be said that there is a probability of 0.5 of actually completing the project before September 14, and a probability of 0.5 of actually completing the project after September 14.

For the simple example in Figure 24, the variances for the Beta distribution of each activity will be:

ACTIVITY			σ^2
A	$\left(\frac{6-2}{6}\right)^2$	=	0.444
B	$\left(\frac{15-5}{6}\right)^2$	=	2.778
C	$\left(\frac{35-13}{6}\right)^2$	=	13.444
Total			16.666

The variance for the total distribution, which is a normal probability distribution, is the sum of the three individual variances, or 16.666. The standard deviation, σ, of the total distribution is then

$$\text{standard deviation} = \sigma = \sqrt{\sigma^2} = \sqrt{16.666} = 4.08 \text{ days}$$

Figure 26, like Figure 25d, shows the total probability curve, with the addition of the standard deviations.

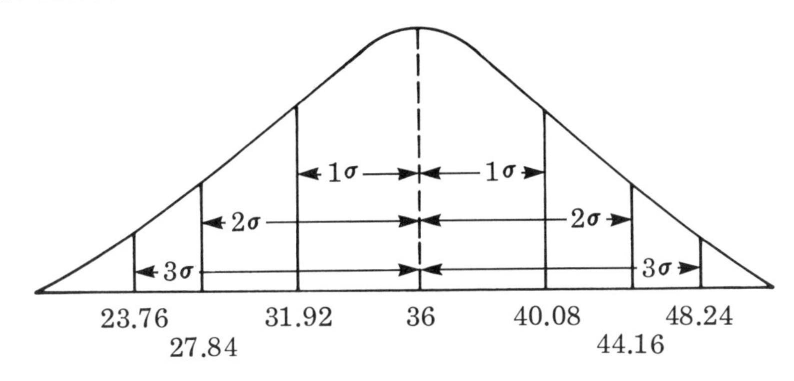

Figure 26.

Figure 26 is a normal curve with 68% of its total area contained within $\pm 1\sigma$ of t_e, or between 31.92 days and 40.08 days. Ninety-five percent of its area is between 27.84 days and 44.16 days, while 99% of its area is between 23.76 days and 48.24 days. As will be seen later, this can be interpreted as:

having a 68% chance of completing the project between 31.92 and 40.08 days;

a 34% chance of completing the project between 31.92 and 36 days;

a 0.95 probability of completing it within 27.84 to 44.16 days;

a 0.475 probability of finishing between 27.84 and 36 days;

a 0.135 probability or 13.5% chance of finishing between 40.08 and 44.16 days;

a 1% chance of either finishing before 23.76 days or after 48.24 days;

a 0.5% chance of the project taking longer than 48.24 days.

Thus, it can be stated that the proportion of the area under certain parts of the normal curve to the total area under the curve is related to probability.

Calculating Probability

The earliest expected occurrence time for the network finish event is determined by the most critical path through the network and is equal to the scheduled time of the network start event plus the sum of the expected durations of the activities on the most critical path leading from the network start event to the network finish event. As stated previously, the probability of actually completing a project before the ET of its network finish event is 0.5, since half of the area under the normal distribution curve is to the left of this ET; and the probability of actually completing the project after the ET of its network finish event is also 0.5, since half of the area under the normal curve is to the right of the ET of the network finish event. Knowing the scheduled time for the network finish event, which is also its LT, makes it possible to calculate the probability of actually completing the project before the scheduled time.

In order to find the probability of actually completing the project before the scheduled time of the network finish event, the following formula is used:

$$Z = \frac{ST - ET}{\sigma_t} \tag{7}$$

The elements in this formula are:

ST is the scheduled time for the event.

ET is the earliest expected occurrence time for the event (mean of normal distribution).

σ_t is the standard deviation of the total distribution of the activities on the longest path leading to the event.

The Z value in the above equation is a measure of the number of standard deviations in the area between ET and ST on the normal probability curve. This Z value then must be converted into a figure that gives the area under the normal curve which lies between ET and ST. Since the total area under a normal curve is equal to 1.0, the probability of finishing the project before ST is equal to the proportion of the area under the curve that is to the left of ST.

An ET of 36 days was calculated for the network finish event for the simple three-activity network in Figure 24 on page 50. The scheduled time for that event is 42 days, or 6 days later than the ET. Figure 27 shows the normal curve for the project, with ET being 36 days and ST being 42 days.

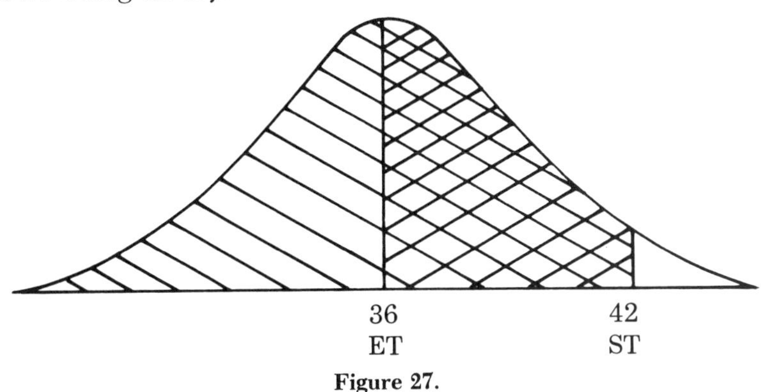

Figure 27.

The proportion of the area under the curve to the left of ST is equal to the probability of finishing the project before 42 days. ET divides the curve into two equal parts, each containing half of the area, so we know that the proportion of the area to the left of ET is 0.5. We must now find the portion of the area between ET and ST, then add this to 0.5 to obtain the portion of the area to the left of ST. Using Equation (7) on page 55 to find the portion of the area between ET and ST:

$$Z = \frac{\text{ST} - \text{ET}}{\sigma_t} = \frac{42 - 36}{4.08} = \frac{6}{4.08} = 1.47$$

The Z value of 1.47 shows that there are 1.47 standard deviations (since in this example, 1 standard deviation is equal to 4.08 days) between ET and ST. However, the Z value does not directly show the proportion of the area under the curve between ET and ST. In order to find this area, the Z value must be converted to a figure that gives the area directly, by using the conversion table on page 57. The first column and the top row are used to find the desired Z value with a significance of 0.01. To find the area for a Z value of 1.47, first look down the leftmost column to 1.4, then go across this row to the 0.07 column. That figure is 42922. This means that for a Z of 1.47 the portion of the area under a normal curve is 0.42922. This figure tells us that the probability of actually completing the project between ET and ST, or between 36 days and 42 days, is 0.42922, or 42.922% chance. However, since we are interested in finding the probability of actually completing the project any time before 42 days, we must now add the probability of finishing before 36 days. Thus, the probability of actually finishing the project before 42 days is equal to the probability of finishing before 36 days plus the probability of finishing between 36 days and 42 days, or:

$$0.50000 + 0.42922 = 0.92922$$

Therefore, the probability of actually completing the project before the scheduled time of 42 days is 0.92922, or 92.922%.

Consider our more complex example of building an outhouse and find the probability of actually completing the project before the scheduled date of the network finish event. The most critical path is 1-2-4-5-6-7-8-9-11-12-13-14-16-17, with an earliest expected occurrence time of September 14 for event 17 and an ST of August 31. Although we already have determined the expected duration for each activity on the most critical path, we must now find the variance for each of these activities in order to obtain σ_t for use in Equation (7) on page 55. The variances for the Beta probability distributions of the activities on the most critical path are shown in Figure 28. These variances, as well as the expected durations for these activities, are

TABLE OF AREAS OF THE NORMAL CURVE BETWEEN THE MAXIMUM ORDINATE AND VALUES OF z (FACTOR 10^{-5} OMITTED)

z	0.00	0.01	0.02	0.03	0.04	0.05	0.06	0.07	0.08	0.09
0.0	00000	00399	00798	01197	01595	01994	02392	02790	03188	03586
0.1	03983	04380	04776	05172	05567	05962	06356	06749	07142	07535
0.2	07926	08317	08706	09095	09483	09871	10257	10642	11026	11409
0.3	11791	12172	12552	12930	13307	13683	14058	14431	14803	15173
0.4	15542	15910	16276	16640	17003	17364	17724	18082	18439	18793
0.5	19146	19497	19847	20194	20540	20884	21226	21566	21904	22240
0.6	22575	22907	23237	23565	23891	24215	24537	24857	25175	25490
0.7	25804	26115	26424	26730	27035	27337	27637	27935	28230	28524
0.8	28814	29103	29389	29673	29955	30234	30511	30785	31057	31327
0.9	31594	31859	32121	32381	32639	32894	33147	33398	33646	33891
1.0	34134	34375	34614	34850	35083	35314	35543	35769	35993	36214
1.1	36433	36650	36864	37076	37286	37493	37698	37900	38100	38298
1.2	38493	38686	38877	39065	39251	39435	39617	39796	39973	40147
1.3	40320	40490	40658	40824	40988	41149	41309	41466	41621	41774
1.4	41924	42073	42220	42364	42507	42647	42786	42922	43056	43189
1.5	43319	43448	43574	43699	43822	43943	44062	44179	44295	44408
1.6	44520	44630	44738	44845	44950	45053	45154	45254	45352	45449
1.7	45543	45637	45728	45818	45907	45994	46080	46164	46246	46327
1.8	46407	46485	46562	46638	46712	46784	46856	46926	46995	47062
1.9	47128	47193	47257	47320	47381	47441	47500	47558	47615	47670
2.0	47725	47778	47831	47882	47932	47982	48030	48077	48124	48169
2.1	48214	48257	48300	48341	48382	48422	48461	48500	48537	48574
2.2	48610	48645	48679	48713	48745	48778	48809	48840	48870	48899
2.3	48928	48956	48983	49010	49036	49061	49086	49111	49134	49158
2.4	49180	49202	49224	49245	49266	49286	49305	49324	49343	49361
2.5	49377	49396	49413	49430	49446	49461	49477	49492	49506	49520
2.6	49534	49547	49560	49573	49585	49598	49609	49621	49632	49643
2.7	49653	49664	49674	49683	49693	49702	49711	49720	49728	49736
2.8	49744	49752	49760	49767	49774	49781	49788	49795	49801	49807
2.9	49813	49819	49825	49831	49836	49841	49846	49851	49856	49861
3.0	49865	49869	49874	49878	49882	49886	49889	49893	49897	49900
3.1	49903	49906	49910	49913	49916	49918	49921	49924	49926	49929
3.2	49931	49934	49936	49938	49940	49942	49944	49946	49948	49950
3.3	49952	49953	49955	49957	49958	49960	49961	49962	49964	49965
3.4	49966	49968	49969	49970	49971	49972	49973	49974	49975	49976
3.5	49977	49978	49978	49979	49980	49981	49981	49982	49983	49983
3.6	49984	49985	49985	49986	49986	49987	49987	49988	49988	49989
3.7	49989	49990	49990	49990	49991	49991	49992	49992	49992	49992
3.8	49993	49993	49993	49994	49994	49994	49994	49995	49995	49995
3.9	49995	49995	49996	49996	49996	49996	49996	49996	49997	49997
4.0	49997	49997	49997	49997	49997	49997	49998	49998	49998	49998

$Z + 10^{-5}$

totaled to obtain the variance and expected duration for the total normal probability distribution. The total variance is 35.694, while the total expected duration is 65 working days. Having the total variance, we must obtain the total standard deviation, σ_t, for use in Equation (7) on page 55. This is found by the following calculation:

$$\sigma_t = \sqrt{\frac{1285}{36}} = \sqrt{35.694} = 5.975 \text{ working days}$$

ACTIVITY	σ^2	t_e
1-2	$\left(\frac{20 - 10}{6}\right)^2 = \frac{100}{36}$	15
2-4	$\left(\frac{20 - 7}{6}\right)^2 = \frac{169}{36}$	11
4-5	$\left(\frac{15 - 2}{6}\right)^2 = \frac{169}{36}$	5
5-6	$\left(\frac{5 - 2}{6}\right)^2 = \frac{9}{36}$	4
6-7	$\left(\frac{3 - 1}{6}\right)^2 = \frac{4}{36}$	1
7-8	$\left(\frac{4 - 1}{6}\right)^2 = \frac{9}{36}$	2
8-9	$\left(\frac{8 - 4}{6}\right)^2 = \frac{16}{36}$	5
9-11	$0 = 0$	0
11-12	$\left(\frac{2 - 1}{6}\right)^2 = \frac{1}{36}$	2
12-13	$\left(\frac{3 - 1}{6}\right)^2 = \frac{4}{36}$	2
13-14	$\left(\frac{5 - 1}{6}\right)^2 = \frac{16}{36}$	4
14-16	$\left(\frac{5 - 3}{6}\right)^2 = \frac{4}{36}$	5
16-17	$\left(\frac{30 - 2}{6}\right)^2 = \frac{784}{36}$	9
Total	$\sigma_t^2 = \frac{1285}{36} = 35.694$	65 working days

Figure 28.

We know the total expected duration for the most critical path is 65 work days. This is equivalent to September 14, which is the earliest expected occurrence time, ET, of the network finish event. We know that the scheduled time, ST, for the network finish event is August 31. Knowing that for a network finish event ST and LT are equal, and also knowing that the event slack for event 17 is −9 work days, it becomes obvious that ST for Equation (7) on page 55 will be 56 work days, which is 9 work days before ET.

Now, using Equation (7), the Z value is

$$Z = \frac{ST - ET}{\sigma_t} = \frac{56 - 65}{5.975} = \frac{-9}{5.975} = -1.51$$

When looking at the Z table on page 57, ignore the minus sign on the Z value and look for the equivalent area for a Z value of 1.51. The equivalent area is 0.43448. Thus, the probability of completing the project between 56 working days and 65 working days, or between August 31 and September 14, is 0.43448.

However, we are interested in the probability of actually completing the project before August 31, or in other words, we want to know the proportion of the area under the curve in Figure 29 that is to the left of ST.

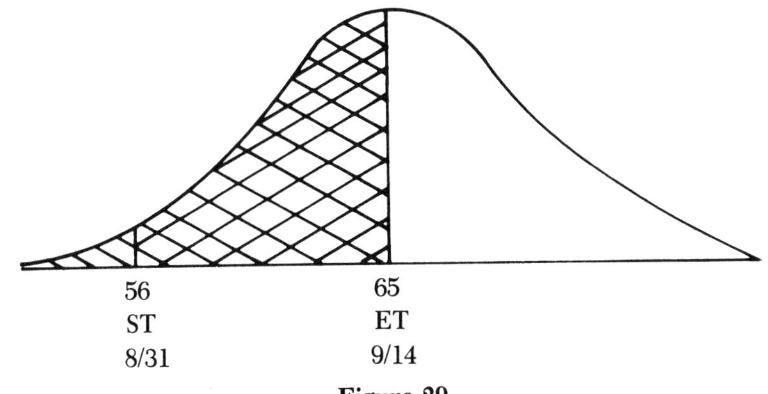

Figure 29.

Since we know the portion of the area between ST and ET to be 0.43448, and we also know the total portion of the area to the left of ET to be 0.5, we can find the portion of the area to the left of ST by this calculation:

$$0.50000 - 0.43448 = 0.06552$$

Therefore, there is a 0.06552 probability, or a 6.55% chance, that the outhouse will actually be completed before August 31.

There is no reason why Equation (7) on page 55 cannot also be applied to milestone events that have scheduled times. In the outhouse example, event 13 was designated a milestone and was given a scheduled time of August 14. The longest path to event 13 is 1-2-4-5-6-7-8-9-11-12-13. Its expected elapsed time is 47 work days, which is equivalent to August 18, which is the ET for event 13. The total variance for this path is 13.36, and, therefore, the total standard deviation, σ_t, is 3.655 days. The scheduled time, August 14, is 2 work days before August 18, so ST is 45 work days. Using Equation (7) on page 55:

$$Z = \frac{ST - ET}{\sigma_t} = \frac{45 - 47}{3.655} = \frac{-2}{3.655} = -0.55$$

Since the equivalent portion of the area under the normal curve for this Z value is 0.20884, there is a 0.20884 probability that event will actually occur between August 14 and August 18 and a 0.29116 probability, or 29.1% chance, that event 13 will actually occur before its scheduled time of August 14.

From the probability calculations made on the examples in Figure 24 and Appendix C, two points should have been noticed:

A. When using equation (7) on page 55 to calculate the probability *for a network finish event*, the numerator, ST − ET, is equal to the event slack, since ST and LT are equal for a network finish event.

B. When using Equation (7) on page 55, if the numerator is positive, then the Z value will also be positive and the probability of the given event actually occurring before its scheduled time will be greater than 0.5. If the numerator is negative, then the Z value will also be negative and the probability of the given event actually occurring before its scheduled time will be less than 0.5.

Conclusion

If each activity in the network has three time estimates, it is possible to calculate the probability of actually completing the project before the scheduled time of the network finish event. It is also possible to calculate the probability of a milestone event actually occurring before its scheduled time.

One should be careful in interpreting probability, since there may be discrepancies that may result from the fact that there may be several paths, other than the longest path, to a given event which are nearly as long as the most time-consuming path. These alternate paths may have substantially different standard deviations than the longest path. If these alternate paths are used for the probability calculations, the probability of the given event actually occurring before its scheduled time may be less than the probability calculated for the longest path. This discrepancy usually arises only when there are two or more paths that are equal or nearly equal in length leading to the event being considered.

Questions

1. In order to calculate probabilities in a network, it is necessary to have three time estimates for each activity (except dummy activities) and scheduled times for network start and finish events. True or False?

2. What is the expected duration, variance, and standard deviation for an activity whose three time estimates are $t_o = 2$, $t_m = 14$, $t_p = 14$.

3. Which of the following is not a measure of dispersion or spread of a distribution: (a) variance, (b) mean, or (c) standard deviation?

4. If the earliest expected occurrence time for a network finish event is 5 days before its scheduled time, what is the probability of completing the project before its scheduled time if $\sigma_t = 4$?

CHAPTER 5

Control: Analyze, Replan, and Update

The controlling function is the process of regulating or directing the progress of a project so that it will be completed on or before the scheduled time of the network finish event. After the initial network plan and schedule have been developed, it is necessary to update the plan and schedule periodically throughout the life of the project so that

1. Actual progress can be compared to the schedule.
2. The network can be replanned if actual progress is behind schedule.

This periodic updating includes:

1. Reporting these actual times:
   a. actual times, AT, that the network start events occurred, and
   b. actual finish times, AF, of activities that have been completed.
2. Analyzing the schedule to determine the status of the project.
3. Replanning the network that includes:
   a. making decisions as how to best eliminate any negative total slack that may exist;
   b. revising the network logic by adding new activities, deleting activities, and/or resequencing existing activities;

c. revising any activity expected durations if more accurate data or new information becomes available; and

d. changing the scheduled times for network start events, network finish events, and/or milestone events.

4. Generating an updated schedule showing the effects of actual progress and the replanning.

QUESTION 23. What are two reasons for periodically updating the network plan and schedule?

The frequency of reviewing and updating the project plan and schedule will vary according to the type, size, and complexity of the project. For example, a project in which the time estimates are given in hours and whose total project duration is expected to be a few weeks may be updated daily; on the other hand, a project for which the time estimates are in weeks and whose total project duration is expected to be several years may be reviewed and updated monthly.

Effects of Actual Times

It is necessary to report actual occurrence times, AT, for the network start events, and actual finish times, AF, for completed activities, to determine the actual progress of the project. This actual progress will have an effect on the schedule of the remaining uncompleted activities in the network, since as the project progresses, some activities actually will be completed before their earliest finish time, some may actually be completed exactly on their EF, and others actually later than their EF. A new schedule, taking into account the actual times of the events that have occurred and the activities that have been completed, can be calculated for the uncompleted activities. These actual times, AT and AF, will have an effect on the calculations of the earliest start and finish times, the slack, and the probability of the remaining uncompleted portions of the network. Looking at the outhouse project, if work on the outhouse project did not actually start until June 18 (rather than June 15), then the actual occurrence time, AT, for event 1, the network start event, would be June 18, and the EF dates for activities 1-2, 1-3, and 1-16 would be 15, 6, and 10 working days beyond June 18, respectively. The earliest start and finish times for all succeeding activities would also change.

QUESTION 24. What would be the EF dates for activities 1-2, 1-3, and 1-16 if event 1 actually occurred on June 18?

If activity 1-2 were actually completed on July 2, the ES for activity 2-4 would now be July 3 and its EF would be July 17. If the ES and EF dates change as the project progresses, then the slack and probability will also be affected.

The network LS and LF times will not be affected by the actual times; they will change only if the scheduled time of the network finish event is changed.

It is a good idea to indicate on the network diagram those portions of the project which have been completed. For example, completed activities can be crossed out, and events that have all their incoming activities completed can be darkened in as shown in Figure 30.

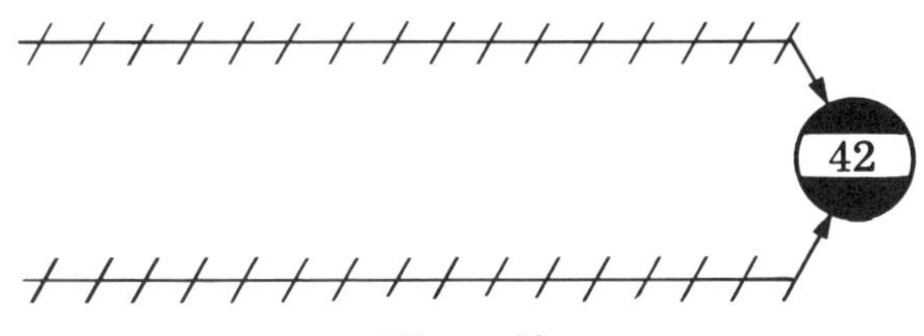

Figure 30.

Analyzing the Network

Network analysis is primarily concerned with investigating the schedule to determine the status of the project, the critical areas or activities in the network, and the causes or reasons for a change (slippage or acceleration) in the schedule as compared to the schedule from the previous updated period.

Checking the earliest expected occurrence time for the network finish event can give a quick indication of the project status. By comparing this ET to the ST for the network finish event, it is easy to see if the project is ahead of or behind schedule.

The critical areas in the schedule can be determined by checking the schedule to find the paths in the network that have negative total slack. It is to these negative slack paths that a concentrated effort must be applied to accelerate the project progress, and bring the project back on schedule. The amount of total slack should determine the priority of applying these concentrated efforts; for example, the most critical path (the one with the most negative total slack value) should be given top priority, while paths with large amounts of positive total slack can be given less attention.

As the project progresses and activities are being completed, various paths in the network will have an incomplete activity at their beginning, and this incomplete activity for a particular path can be designated "could be in progress," since all of its predecessor activities will have been completed, and nothing is stopping it from being performed. Attention should be focused on expediting the completion of these "in-progress" activities at the beginning of the critical paths (negative total slack), since network logic dictates that the succeeding activities on these paths cannot even start until the "in-progress" activities are completed. Priority should be given to accomplishing the in-progress activities that have large amounts of negative total slack rather than to those activities that have large amounts of positive slack.

QUESTION 25. To accelerate project progress, to which paths should a concentrated effort be applied?

The most difficult part of network analysis is determining the causes or reasons for changes in the schedule from the previous updated schedule. This is a difficult task because the change in the status of a project from one update period to the next may be due to a complex combination of factors. From one update period to the next, many activities may have been completed, the network logic may have been changed by adding, deleting, or resequencing activities, and some activity durations and event scheduled times may have been revised. However, it is very important to determine the causes of a change in the schedule so that effective control can be exercised in replanning the remaining uncompleted portions of the project.

Replanning the Network

Once the network has been analyzed to determine what areas of the project are critical, it is necessary to decide exactly how to revise the plan to eliminate these critical areas and also to include any new information.

First consideration should be given to what replanning is necessary to help reduce the negative total slack in the network. There are several approaches that can be used to reduce the negative slack, and they mostly have to do with reducing the expected duration of one or more of the activities on the critical paths. It must be remembered that total slack for a path is shared by all the activities on that path, and a change in the expected duration of any of the activities that form that path will also cause a proportionate change in the total slack of that path.

One approach used to reduce the negative total slack of a critical path is to check if any of the activities on the path can be performed in parallel, rather than in series, thus reducing the total elapsed time for that particular path.

A second approach involves reducing the expected duration of one or more activities on a critical path by reducing the specifications or requirements for an activity to be considered complete. Looking at our example, the t_e of activity 13-14, painting the interior, is 4 work days. If this may be based on using two coats of paint, the t_e of this activity may be reduced by using only one coat of paint. To carry this approach to the extreme, the project plan may be reevaluated to determine if any critical path activities can be eliminated entirely.

Another approach, and probably the most obvious, is to reduce the expected durations of certain activities on critical paths by using additional resources, either equipment or personnel. For example, the t_e for activity 2-4, clearing land, may be reduced by using more people.

Another obvious approach is to reduce certain activity expected durations by using automated equipment. For example, the t_e of activity 13-14, painting the interior, may be shortened if a paint spray gun is used for painting rather than brushes.

Still, another obvious approach to reducing the expected durations of certain activities is by working overtime, by increasing the work hours per day or the work days per week, or both.

If it is impossible to reduce any of the expected durations for any of the activities on critical paths, then it may become necessary to choose a different and later scheduled time for the network finish event.

Besides replanning the network to reduce the negative total slack of the critical paths, other replanning also may take place by the introduction of new information. Activities that may have been forgotten when the original plan was developed may be added, or a new segment may be added to the project. Replanning also may involve deleting some activities that are no longer required due to a change in some portions of the project. It also may mean resequencing existing activities to change the logic of certain parts of the network. This addition, deletion, and change in the network logic may be the result of various factors as the project progresses.

Replanning also may require revising the duration estimates for some activities because of the availability of more accurate information. For example, assume a network has ten *similar* activities, eight at the beginning of the project and two near the end of the project. After the first eight activities are completed and their actual elapsed durations are known, the duration estimates for the two similar activities near the end of the project may be revised. The replanning procedure also can mean changing or revising any of the scheduled times in the network, whether it be for network start events, network finish events, or milestone events.

After replanning has been completed, it is then possible to calculate a revised, or updated, schedule to show the effects of actual times and the effects of replanning.

Updating the Schedule

The new information from the replanning as well as the actual times of completed portions of the project are used to calculate an updated schedule for the remaining uncompleted portions of the project. In this new schedule ES, EF, LS, LF, ET, LT, slack, and probabilities may change slightly or drastically for some or all parts of the project. For example, a path that originally may have been the most critical path, may, several weeks or months after the project has started, become less critical, and a different path may then become the most critical path.

In order to show an updated or revised schedule, let us consider the outhouse example. Assume that on July 23 it is decided to review the project and produce an updated schedule using the following available information:

1. Actual time for event 1 was June 18.
2. Actual finish times for completed activities:

1-2	July 2
1-3	July 8

2-4	July 13
3-4	July 16
4-5	July 21
4-12	July 23
5-10	July 22

3. Reduce the expected duration of activity 5-6, build frame, from 4 to 3 work days by having a visiting relative help with the work.
4. Change the expected duration of activity 4-13, clear path from the main house, from 7 to 9 days due to encountering unexpected problems.
5. Change the expected duration of activity 8-9, put up the interior walls, from 5 days to 3 days by working overtime.
6. Add two new activities: 7-18, install vent pipe, and 18-8, shingle roof, with time estimates (t_o, t_m, t_p) of 1-2-3 and 1-1-2 work days, respectively.
7. Change the scheduled time for event 17 from August 31 to September 4.

(At this point the reader should refer to Appendix E for reference while reading the rest of this chapter.)

Figure 31 on page 71 shows the updated version of the original schedule calculated in Chapter 3 (see Figure 19 on page 35). Similarly, Figure 32 on page 73 shows the updated schedule of event times compared to the original schedule calculated in Chapter 3 (see Figure 20 on page 37). All calculations are based on the calendar on page 72.

Rule 5. The earliest start time, ES, for an activity or activities leading from an event whose incoming activities have all been completed, is equal to or later than the latest of all the actual finish times of all the completed activities coming into that same event.

The early finish time, EF, for the activity is found by adding its expected duration, t_e, to its ES (EF = ES + t_e). However, if the report date, RT (when the updated schedule is being calculated), is later than the EF for the activity, then substitute RT for EF (and in this case an ES would be meaningless).

The second part of the above rule addresses activities that are overdue for completion, that is, those that should have been completed, EF, by the report date, RT, but were not. To take this lateness into account, RT is substituted for EF; this gives a current and true value for the total slack. For example, in making the calculation for the updated schedule in Figure 31 on page 71, one might expect the ES for activity 1-16, collect newspaper, to be June 18, since that is the AT for the network start event, and the EF to be 10 days later on July 1. Also, if these dates are used as the ES and EF, the total slack calculation would be +38 days for activity 1-16. However, since the report date, RT, of the updated schedule is July 23, it would be foolish to state that the earliest finish time for an activity is July 1, since this date has passed. To give a more realistic indication of the total slack, RT is substituted for EF. In essence, this says that the activity should have been completed on July 1, but it wasn't, and, in fact, as of the report date it still isn't completed. Therefore, a slack of +38 days is not correct, but the slack is reduced to +22 days when RT is substituted for EF; this shows that, as of July 23 (the report date), this activity really has a total slack of +22 days (not 38).

Looking at other activities in the schedule, activity 4-13, clear path, has an ES of July 17, since the AT of event 4 was July 16, which is the later AF of activities 2-4 and 3-4. The EF for 4-13 is July 29, which is 9 work days beyond July 17; activity 5-6 has an ES of July 22, since its preceding activity 4-5 was finished on July 21.

Looking further at the schedule, the LS and LF dates were also changed for the uncompleted activities, since the ST of the network finish event was changed from August 31 to September 4. The most critical path is 5-6-7-18-8-9-11-12-13-14-16-17, with a total slack value of +1 work day. Therefore, most critical total slack has changed by 10 work days, from −9 to +1. This 10-day change in total slack was caused by the extension of the scheduled time for event 17 by 4 work days (August 31 to September 4), the occurrence of event 5 four work days before its ET on the previous schedule (AT is July 21 compared to ET of July 27 in Figure 20), and the shortening of t_e for activities 5-6 and 8-9 by 2 and 1 work days, respectively. This is a gain of 4 + 4 + 2 + 1 = 11 work days, but activities 7-18 and 18-8 replaced activity 7-8 on the most critical path. However, their combined t_e was 2 + 1 = 3 work days, compared to 2 work days for activity 7-8, so the net gain in total slack was 11 − 1 = 10 work days. The probability of actually finishing the project before September 4 is 0.58, compared to the previous schedule which has a 0.07 probability of actually finishing before August 31. Also, the probability of event 13 actually occurring before its scheduled time of August 14 has changed from 0.29 to 0.99.

REPORT DATE: July 23

P.E.	S.E.	ACTIVITY DESCRIPTION	t_e	ES	EF	LS	LF	TS	FS	AT
1	2	Choose Location								7/2
1	3	Decide on Style								7/8
1	16	Collect Newspaper	10		7/23	8/11	8/24	22	21	
2	4	Clear Land								7/13
3	4	Draw Plans								7/16
4	5	Buy Materials								7/21
4	12	Dig Hole								7/23
4	13	Clear Path	9	7/17	7/29	7/30	8/11	9	8	
5	6	Build Frame	3	7/22	7/24	7/23	7/27	1	0	
5	10	Build Door								7/22
6	7	Put On Roof	1	7/27	7/27	7/28	7/28	1	0	
7	8	Put On Exterior Siding	2	7/28	7/29	7/30	7/31	2	1	
7	18	Install Vent Pipe	2	7/28	7/29	7/29	7/30	1	0	
8	9	Put Up Interior Walls	3	7/31	8/4	8/3	8/5	1	0	
9	11	Dummy Activity	0	8/4	8/4	8/5	8/5	1	0	
9	12	Sand Seat	1	8/5	8/5	8/7	8/7	2	1	
10	11	Cut Half Moon in Door	1	7/23	7/23	8/5	8/5	9	8	
11	12	Mount Door	2	8/5	8/6	8/6	8/7	1	0	
12	13	Place Bldg. over Hole	2	8/7	8/10	8/10	8/11	1	0	
13	14	Paint Interior	4	8/11	8/14	8/12	8/17	1	0	
13	16	Paint Exterior	7	8/11	8/19	8/14	8/24	3	2	
14	15	Install Door Latch	2	8/17	8/18	8/21	8/24	4	0	
14	16	Hang Pictures	5	8/17	8/21	8/18	8/24	1	0	
15	16	Dummy Activity	0	8/18	8/18	8/24	8/24	4	3	
16	17	Test	9	8/24	9/3	8/25	9/4	1	0	
18	8	Shingle Roof	1	7/30	7/30	7/31	7/31	1	0	

Figure 31. Activity Schedule

	S	M	T	W	T	F	S
		1	2	3	4	5	6
	7	8	9	10	11	12	13
JUNE	14	15	16	17	18	19	20
	21	22	23	24	25	26	27
	28	29	30				
				1	2	3	4
	5	6	7	8	9	10	11
JULY	12	13	14	15	16	17	18
	19	20	21	22	23	24	25
	26	27	28	29	30	31	
							1
	2	3	4	5	6	7	8
	9	10	11	12	13	14	15
AUGUST	16	17	18	19	20	21	22
	23	24	25	26	27	28	29
	30	31					
			1	2	3	4	5
	6	7	8	9	10	11	12
SEPTEMBER	13	14	15	16	17	18	19
	20	21	22	23	24	25	26
	27	28	29	30			

REPORT DATE: July 23

EVENT	ET	LT	EVENT SLACK	AT	ST	PROB.
1				6/18	6/15	
2				7/2		
3				7/8		
4				7/16		
5				7/21		
6	7/24	7/27	1			
7	7/27	7/28	1			
8	7/30	7/31	1			
9	8/4	8/5	1			
10				7/22		
11	8/4	8/5	1			
12	8/6	8/7	1			
13	8/10	8/11	1		8/14	.99
14	8/14	8/17	1			
15	8/18	8/24	4			
16	8/21	8/24	1			
17	9/3	9/4	1		9/4	.58
18	7/29	7/30	1			

Figure 32. Event Schedule

Conclusion

The key to controlling the project plan is preventing trouble before it starts by making use of the capability of network planning to predict or forecast trouble areas. This may be the reason why network planning seems to be better for preventing negative slack than for eliminating it.

Network planning is a project management tool and will not replace the decision making process. Rather, it aids management by emphasizing the areas where problem solving and decision making are needed, thus allowing for management by exception rather than "seat of the pants" decision making. By analyzing the schedule, the causes of any accelerations or slippages can be determined, and the areas of criticality in the plan can be found. This provides a sound basis for deciding which replanning approaches would be most effective in reducing the critical areas of the project. Slack analysis also helps provide a more efficient use of resources by showing where they can be used to best advantage. For example, it indicates the exact activities on which work must progress if the project is to keep on schedule, and also prevents "spinning wheels" or "putting out fires" in the noncritical areas of the project.

The controlling function is really an analysis–synthesis process whereby first a schedule is analyzed, and then the replanning and actual accomplishments are synthesized into an updated or revised schedule. This continuous updating of the project enables effective control throughout the life of the project.

Appendix F shows the updated project plan for the outhouse example using the activity-in-the-box format.

Questions

1. Why is it necessary to review the network plan periodically during the life of the project?
2. What is: (a) AT, (b) AF.
3. What calculations will be affected by AT and AF?
4. A change in the expected duration of any of the activities on a path in a network will cause a proportionate change in the total slack of that path. True or False?
5. Name several ways of reducing the expected durations of activities.

CHAPTER 6

Resource Considerations

Network scheduling is basically a time-oriented technique. One of the major assumptions used for the schedule calculations in Chapters 3, 4, and 5 is that the resources required for the project are not a limiting factor. In other words, it is assumed that all required resources will be available at any time they are needed.

Resources may consist of people, equipment, tools, space, etc. If there are a limited number of resources available to perform the project activities, then there are two approaches to take this limitation into account: The first and more frequently used approach is one that takes the resource limitations into account when drawing the logical relationships of activities in the network diagram. Network diagrams are developed illustrating, as a minimum, the *technical constraints* among activities. In other words, activities are placed in a series relationship, for example, because from a technical standpoint the activities must be performed that way. An example in building a house would be that the three activities of (1) build foundation, (2) build frame, and (3) put on roof must be done in series, as shown in Figure 33. Technically, these activities must be performed in this sequence. One could not put on the roof first and build the foundation later!

In addition to the network logic showing the technical constraints among activities, the logic could also show *resource constraints* among activities. In other words, activities are drawn in logic relationships that are based on *both* technical and resource constraints. For example, a portion of a project plan for building a house may include the following activities: (1) paint livingroom, (2) paint kitchen, and (3)

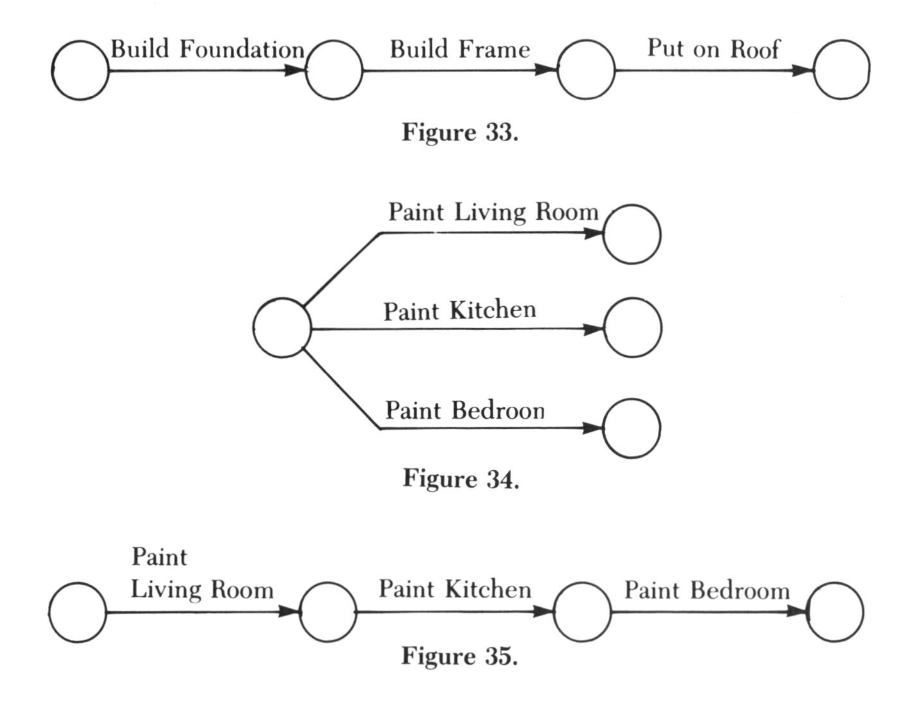

Figure 33.

Figure 34.

Figure 35.

paint bedroom. From a technical viewpoint these three activities could all be performed in parallel, that is, there is no technical reason why the start of any one of these activities should depend on any other being completed. Therefore, they could be drawn as shown in Figure 34.

However, if there is only one person who will be doing all the painting, this factor now introduces a resource constraint. That is, although technically all three activities could be done in parallel if sufficient resources were available, they will have to be performed in series, since only one person will do all three activities. Because of this resource constraint, the logic will have to be drawn as shown in Figure 35. The exact sequence of these three activities in terms of which room gets painted first, second, and last is a personal preference or a separate decision.

This example illustrates how resources may be taken into account by considering the logical relationships among activities in the network. Of course, this approach assumes that the user has some sense of how many resources will be available to perform the project activities and keeps this in mind when developing the network logic.

A second approach to take resources into account is one that is fairly complicated (the calculations are usually done by a computer) and requires the user to provide additional information about how many resources are needed to perform each activity and how many total resources are available during the entire duration of the project. With this method, the network is developed based only on technical constraints among activities; that is, all activities that technically can be done in parallel are drawn that way. Next, in addition to duration estimates for each activity, the user must also estimate the number of each type of resource required for each activity. In addition to knowing how many of each type of resource is needed to perform each type of activity, it is also necessary to know the total number of each type of resource that is available at any time during the life of the project. With all this information, the (computer) calculations are made in such a manner that, for any period of time (day, week, etc.), available resources are assigned to activities (which are scheduled to be performed during that period—based on their ES and EF times) on a priority basis starting with those that have the least value of total slack. This continues until all available resources are assigned. If any activities did not get resources (because all available resources were used up), then these activities are delayed until the next time period. This, in essence, lowers their total slack amount, and they become more critical. Once they are delayed into the next time period, they once again compete, on a priority basis, with other activities scheduled for that time period that require the same type of resources.

This iterative scheduling process continues and may result in the overall project schedule being extended because of insufficient levels of certain types of resources. A decision can then be made on whether additional resources should be made available to reduce the overall project schedule. The calculations involved in the second approach are both complicated and cumbersome and are difficult to do by hand. However, there are project scheduling computer programs that perform such calculations, but they still depend on accurate input data being provided by the user.

When preparing a network diagram, it may be helpful to indicate the name of the person or organization who has the responsibility for accomplishing each activity. This name or code could be written under each arrow (or in each box when using the activity-in-the-box format or on the work breakdown structure chart). This will help one to visualize which resources may be required and may aid in developing the network logic to take resource constraints into account. Similarly, the person or organization responsible for each activity could be listed on the project schedule. Each person then need only check the schedule for the activities for which they are responsible. (There are computer programs that will sort the activity schedules by responsibility codes, which the user assigns. Thus, all the activities for a specific person or organization will be grouped together.)

CHAPTER 7

Computer-Aided Planning and Scheduling

There are many computer programs (software) available to do the scheduling calculations required when using network planning techniques. The computer does not replace the planning, scheduling, or controlling functions; nor does it replace the management decision making process. The computer merely does the arithmetic computations necessary for obtaining a schedule, and presents the schedule information in various report formats which then must be analyzed to make decisions regarding any alterations to the plan.

To Use or Not to Use—the Computer

Although a computer can perform the scheduling calculations much faster and with less error than can be done manually, that alone may not be reason enough to use a computer. The following points should be considered in determining whether or not a computer should be used:

1. *Size of the network.* The number of activities in the network diagram is probably the most obvious consideration. There is no one magic number, such as 100 activities, that is the break point in deciding to use a computer. Rather, it can be said only that the more activities in a network plan, the more reason there is to use a computer.
2. *Complexity of network.* The complexity of a network depends on the number of interrelationships among the activities in the network. Larger networks are not necessarily more complex

than smaller networks. For example, there may be a network of 200 activities that are arranged in four paths of 50 serial activities with little or no interrelationships among the four paths; while another 200-activity network may have a large number of series–parallel interdependent relationships among the activities. In general, the user must decide if the network is complex enough to merit using a computer.

3. *Frequency of updating.* The number of times the project will be reviewed and an updated schedule generated is another consideration. Projects can be a few days long or they may last several years, but regardless of the length of the project, the number of times the plan is reviewed and the schedule is updated throughout the project are important considerations. For example, a project may take five years and be updated once a year, or a project may last one week and be updated every eight hours. In general, the more frequently a project is updated, the more reason there is to use a computer.

A network does not have to be large *and* complex *and* frequently updated to justify using a computer to do the schedule calculations. Rather, any one or a combination of these factors may determine whether to use or not to use a computer.

Computer Program Features

There are many project planning computer programs available from computer manufacturers, computer software companies, or management consulting firms (see Appendix L). These computer programs can range from very simple and straightforward to highly sophisticated and complex. They are available for different sizes of computers: large mainframe computers, minicomputers, and personal or microcomputers.

The following list of features of project planning and control computer programs may act as a guide for comparing various programs. These features are not listed in any particular sequence or order of priority.

1. *Capacity.* Capacity means the maximum number of activities that can be included in the network. Some computer programs can handle several hundred activities, while others can handle thousands.

2. *Event or activity numbering system.* There are three common types of numbering: sequential, random, and alphanumeric.

   a. Sequential event numbering requires that an activity's ending event number be larger than its beginning event number. In other words, each arrow in the network must have a higher event number at its head than at its tail. This requirement may make it difficult for the user to add activities when replanning, without renumbering the entire network.

   b. Random numbering, on the other hand, does not require a larger ending event number than a beginning event number. With random event numbering, activities can be added to the network during replanning stages without concern about having a smaller beginning event number than an ending event number.

   c. The alphanumeric system allows the use of a combination of numbers and alphabetic characters to identify events. For example, the user may want to have various subnetworks identified by a certain alphanumeric code.

3. *Duration estimates.* Some computer programs allow for three duration estimates—optimistic, most likely, and pessimistic—for each activity. Other programs allow for two time estimates—normal and "crash"—and still others permit only one estimate for each activity. Some programs allow for the choice of using one or three, or one or two, duration estimates for each activity.

4. *Activity- and/or event-oriented.* Programs are activity-oriented when they allow use of descriptions for activities and when the output reports give activity-type information, such as start and finish times. Event-oriented computer routines allow only event titles and have output reports with event-type information, such as earliest and latest occurrence times. Still there are other programs that allow the user to input both activity descriptions and event titles, having both activity and event output reports.

5. *Descriptions.* Some programs do not allow the use of activity descriptions and event titles, and therefore the activities listed

on the output reports are identified only by their beginning–ending event number combination. Even the programs that permit description input information vary. Some allow as few as 12 characters to be used for each description, while others allow over 35 characters.

6. *Multi-start and/or multi-finish networks.* Some computer programs allow only one network start event and one network finish event, others allow more than one network start event but only one network finish event, and still other programs permit more than one network start and finish events. If you have a multi-start and/or multi-finish network, but you have a program which allows only one network start and finish event, you must use dummy activities to tie together all the start events into one network start event and all the finish events into one network finish event.

7. *Scheduled event times.* Scheduled times are used in conjunction with network start and finish events and milestone events.

   a. There are computer programs that permit scheduled times to be used only for the network start events, while others permit the use of scheduled times for both network start and finish events.

   b. Some programs permit the use of scheduled times for milestone events throughout the network. Some of these programs do not take these scheduled times into consideration for calculation purposes; rather, they print them on the output reports for informational purposes only. When the scheduled times for milestone events are used in calculations, the program either (1) compares the scheduled time with the event's earliest expected occurrence time, using the later of these two times for forward calculations from that event, or (2) compares the scheduled time with the event's latest allowable occurrence time, using the earlier of these two times in the backward calculations from that event.

8. *Responsibility code.* Some programs permit the user to input a responsibility code for each activity. This code is used to indicate which persons, departments, or organizations are re-

sponsible for accomplishing a particular activity. These responsibility codes cn range from 1 to 18 alphanumeric characters depending on the program.

9. *Time base.* Different computer programs have different time bases. Some may allow time estimates only in days, others in hours, and others in weeks and tenths of weeks. There are some programs that have a variable time base; that is, the user tells the computer what time base to use.

10. *Calendar date output.* There are some programs that show the calculated activity and event times in calendar date format on the output reports. For example, if days or weeks are the time base, the calculated times will be shown in a day–month–year format. Programs that do not have calendar date outputs merely show the calculated times as the total number of hours, days, or weeks from a project start time of zero.

11. *Variable work week.* Programs that have the variable workweek feature permit the user to vary the number of work days per week and/or the number of work hours per day during the project. Thus, a project schedule can be compressed by increasing the number of work hours per day and/or work days per week. Some programs allow the user to designate the workweek for each individual activity in the network, so, for example, some activities may be scheduled only for a 5-day work week, others for a 6-day work week, and still others are allowed to be scheduled for work only on weekends.

12. *Holiday allowance.* Computer programs with this feature permit the user to designate certain holidays during the life of the project when work will not be scheduled. As a result, the program will not consider these days as work days when making the calculations. Some programs permit only a few holidays, while others permit hundreds of holidays throughout the life of the project.

13. *Actual times.* There are actual times for events and actual times for activities.

    a. An actual time for an event is the time that the event actually occurs, that is, when all activities leading into the event have been completed. The actual event time can also be the actual starting time of a network start event, which may differ from its scheduled start time.

b. There are two types of actual times for activities: (1) actual start times and (2) actual completion times.

Some programs do not take the event and activity actual times into consideration when making calculations; instead, the actual times are displayed in the output reports for informational purposes only. However, there are computer programs that do take these actual times into consideration when making the forward calculations through the network to obtain the earliest times, generating a revised schedule based on the actual times that events have occurred and the actual times that activities have been started and completed.

14. *Updating*. There are programs that make calculations based on the first network diagram, but then do not allow for the updating of the schedule as the project progresses. In other words, these programs do not allow for the addition, deletion, or rearrangement of activities, nor do they consider actual times. These programs should be used only to develop an initial schedule for a project for which replanning is not expected.

    On the other hand, there are computer routines that allow the user to update various changes in the network logic, activity time estimates, and scheduled times and also permit the user to input actual times as the project progresses. With these new data, an updated schedule is produced.

15. *Partial progress*. Some computer programs allow the user to input partial completion of activities by either (1) indicating the percentage completion of an activity or (2) indicating the remaining duration to complete an activity. This information is taken into account when calculating an updated schedule.

16. *Data input*. The degree of "user-friendliness" of the computer program is a feature to be considered. Some programs allow for easy data entry by incorporating such features as menus, prompting, touch-screen control, and HELP functions. With these features, network data can be entered into the computer faster and with fewer errors.

17. *Slack or float.* There are various types of slack or float: total, free, interfering, independent, scheduled, and event slack. Although most programs calculate total slack, some will also calculate some of the other types of slack or float.

18. *Negative total slack.* Not all programs permit negative total slack. The programs that do not allow scheduled times for the network finish events usually cannot calculate negative slack values, since there is no scheduled time from which to make the backward or latest time calculations. These programs merely use the forward-calculated earliest expected occurrence time of the network finish event as reference for backward calculations. In other words, it makes the latest time equal to the earliest expected occurrence time for all network finish events. As a result, the most critical path will have zero total slack and the total slack will never become negative. However, programs that allow the user to specify scheduled times for the network finish events usually calculate negative total slack.

19. *Slack allocation.* There are some computer programs that allow the total slack for a particular path to be allocated among the activities which are on that path. Other than the manner in which it is normally calculated for activities, slack can be allocated in the following ways:

    a. A slack weight factor is permitted to be assigned to each activity in the network. For example, the slack weight factor may be a number from zero to nine. The total slack for a given path is then allocated among the activities so that the activities which have the higher slack weight factors are assigned a greater proportion of the total slack for that path if the total slack is positive, and proportionately less if the total slack is negative. Some computer programs work in the opposite manner; that is, the activities with lower slack weight factors are assigned greater proportions of positive total slack and less proportions of negative total slack.

    b. Another method of allocating slack is to assign the activities near the end of a given path proportionately more of the total slack for that path than the activities at the beginning of the path if the total slack has a positive value. If the total slack for a path is negative, the activities near the end of the path are assigned proportionately less of the negative total slack for the path than the activities at the beginning of the path.

c. A third manner of allocating slack is to specify a certain amount of slack for any of the activities in the network. If there is any positive total slack for a path, it is assigned to these activities according to the amount specified by the user; the remaining slack is then shared equally among the other activities on the path.

d. A final method of allocating total slack among activities on a particular path is by allocating positive total slack among the activities in direct proportion to their expected duration, and allocating negative total slack among the activities in an indirect or inverse proportion to their expected duration.

Computer programs that have the slack allocation feature may permit the total slack to be allocated by one or several of the methods described.

20. *Error diagnostics.* Some computer programs have the ability to check the input data for various kinds of errors. Some error detection involves checking individual activities and events. For example, it may check if the user has erroneously input an alphabetic character for the time estimate or if the pessimistic time estimate is less than the optimistic time estimate for a particular activity.

Other error diagnostics involve checking the entire network for network-type errors, such as loops, completed activities with incomplete predecessor activities, or network start and finish events lacking scheduled or actual times.

21. *Statistical analysis.* For the computer to make any probability calculations the user must input three duration estimates for each activity, except dummy activities. Most computer programs that make statistical computations show the probability of making scheduled times for network finish events and milestone events.

Some of these programs also calculate either the total variance or standard deviation of the normal total probability distribution for the scheduled event times; and others calculate either the variance or standard deviation of the Beta probability distribution for each activity in the network.

22. *Output report sorting.* Some programs do not give the user a choice of how the schedule or output reports are sorted. For example, the program may automatically sort all activities by beginning event number; other computer routines do give the user a limited or wide range of choices of sorting the schedule information. Common sorting keys are beginning event number, ending event number, responsibility code, earliest start time, earliest finish time, latest start time, latest finish time, or total slack, for activity reports. For event reports, the common sorting keys are event number, earliest expected occurrence time, latest allowable occurrence time, or event slack.

    There are some programs that allow the user to sort the schedule on a major sorting key and one or more subsorting levels. For example, the user may want to sort first by responsibility code, then subsort all the activities for each different responsibility code by the latest finish time, and if more than one activity under each responsibility code has the same latest finish time, then the activities would be further sorted by total slack (from most negative to most positive). In other words, the major sort would be the responsibility code, the first level subsort would be the latest finish time, and the second level subsort would be total slack.

23. *Report date.* This feature allows the user to input to the computer the "as of" date of the schedule. The computer program may use this report date for calculations in two ways:

    a. It compares the scheduled dates of all the network start events with the report date, and if the report date is later than the scheduled date of any of the network start events, then the report date is substituted for the scheduled date (since the scheduled date is overdue) for those particular network start events and is used for making the forward calculations through the network.

b. It compares the earliest finish dates of the activities (which have all their predecessor activities completed) with the report date, and if the report date is later than the earliest finish date of any of these activities, then the report date is substituted for the earliest finish date (since the EF is overdue) for those particular activities and is used for making the forward calculations through the network.

This report date feature allows the total slack to be kept current by allowing the slack to become smaller if work has not been performed on activities that should be in progress (all predecessor activities completed) and also on those overdue for completion (the activity's earliest finish date is earlier than the report date).

24. *Flags.* Some computer programs have various flags or indicators to call the user's attention to certain parts of the schedule. For example, an asterisk may appear alongside each activity on the most critical path, or an asterisk will appear next to the earliest start, earliest finish, latest start, and latest finish dates for each activity as each date is surpassed by the report date. This tells the user that the activity "should be in progress" if the start dates are flagged or that the activity is "overdue for completion" or "late" if the finish dates are flagged.

The asterisk is the most commonly used flag, although other characters may be used.

25. *Condensed output reports.* This feature allows the user to condense the schedule reports by suppressing unwanted information. For example, in order to shorten output reports as the project progresses, the user may not want completed activities with their associated actual finish times to appear on the schedule reports. Or the user may want to see only those activities with negative slack or only those activities with certain responsibility codes.

26. *Network summarization.* This feature allows large networks to be reduced to varying degrees into summary networks. Network summarization gives information only on certain key or milestone events rather than for all events. For example, for

a network with 2,000 events, a level C summary may give the schedule for only 500 events, a level B summary may produce a schedule for only 100 events, and a level A summary may produce schedule information on the 40 most important events in the network. When the user asks for a level B summary report, schedule information will be produced for all 40 A level events and the 60 events that were designated as B level events.

Network summarization is particularly useful for reporting schedule information to various levels of management. Computer programs vary as to how many levels are permitted. Some allow only one summary level, while others allow over 25 different levels of summarization. When using this feature, the user must designate a level for each event in the network. Events not given a level will not be considered in any summary reports and will only appear on detailed reports. All network start and finish events as well as any milestone events with scheduled times should be designated as the highest summary level.

27. *Graphical output.* Some computer programs have the capability of producing graphical output in the form of either a network diagram or a bar (Gantt) chart. Some people prefer a bar chart output because they feel it presents a better visual display of when each activity must take place. Some programs will also time-scale the network plots or have features that will "zone" the plot by responsibility code or will produce a plot for a particular "window" of time rather than for the entire project duration.

28. *Cost summarization.* This feature allows the user to input an associated cost estimate for each activity. The program then adds all the costs and gives the total planned costs of the project. Other programs also allow the user to input the actual costs for completed activities. These programs sum up both the planned and actual costs for the entire project.

29. *Cost forecasting.* Some programs go beyond cost summarization and use the actual cost data to forecast what the total actual costs will be for the entire project. These computer programs may forecast total actual costs in one of two ways.

a. The computer program compares actual costs to planned costs for completed activities and calculates the rate of spending. The program assumes that the user will continue to spend money at this same rate for the remainder of the project, and based on this assumption it calculates a forecasted total cost at completion for the entire project.

For example, the summation of the planned costs of all the activities in a project sums to $10,000. To date $1,500 actual costs have been incurred for completed activities; the planned costs for these same completed activities are $2,000. The computer calculates that the user is spending at a rate of 1,500/2,000 or 0.75. The program will now assume that the user will continue to spend at a rate of .75, and it will forecast a total cost at completion of 0.75 (10,000) or $7,500 for the entire project.

b. Another method of forecasting total actual costs assumes that regardless of how actual costs have compared with planned costs for completed activities, the user will continue to spend an amount equal to the planned costs of the remaining incomplete activities. In other words, it assumes that for the remainder of the project the user will spend the same amount originally estimated for those incomplete activities. So the program calculates a forecasted total cost at completion by adding the actual costs for the completed activities and the planned costs for the incomplete activities.

For example, the total of the planned costs of all the activities in a project total $10,000. To date the user has incurred $1,500 actual costs for completed activities; the planned costs for these same completed activities are $2,000. The program will assume that the user will spend $8,000 ($10,000–$2,000) on the remaining portion of the project. It will add this $8,000 planned costs for the incomplete activities to the $1,500 actually spend on the completed activities and forecast a total cost at completion of $9,500 for the entire project.

30. *Cost optimization.* When a program has this feature, the user must input both "normal" and "crash" time estimates and ac-

companying "normal" and "crash" costs for each activity. The "crash" time is less than the "normal" time, and the cost of accomplishing the activity in "crash" time is greater than the cost of "normal" time. The computer program then "trades off" between time and cost and produces not one schedule but a set of schedules. These schedules vary in total project duration and costs. The shorter the total project duration, the higher the cost, but for any one particular schedule, the cost is the optimum cost for that total project duration. Optimization is accomplished by "crashing" those critical path activities that have the smallest cost increase for the associated shortening of the total project duration.

31. *Resource summarization.* Computer programs that have this feature permit the user to input the amounts of various types of resources that are needed to accomplish each activity. Resources may be various types of personnel (carpenters, drivers, programmers, etc.) or equipment (forklifts, testing stations). Some programs allow only one type of resource per activity, while others allow many different types of resources. The program then totals the resources and gives the total number of each type of resource that is needed to complete the project.

    Some programs with this feature also produce a total resource cost for each activity or for the entire project, if the user specifies the per unit cost rate for each type of resource. The program merely multiplies the number of each resource by its per unit cost rate and gives the total resource cost. It should be noted that these programs assume a direct linear relationship between total costs and the total number of resources for each type of resource.

32. *Resource leveling (smoothing).* This feature attempts to minimize both overtime and waiting or stand-by time for resources within a fixed total project duration. There are three types of resource leveling: (1) fixed leveling, which limits the user to a fixed number of each resource for the life of the project; (2) variable leveling, which allows the user to increase or decrease resources as needed throughout the life of the project; and (3) a combination of fixed and variable leveling, which gives the

user a minimum fixed number of resources and allows for an increase of additional resources above the minimum fixed number. Depending on the program, it may permit one or more of these types of leveling. Programs will also vary as to the number of different types of resources that can be used.

33. *Resource allocation.* Given a fixed number of resources, computer programs with resource allocation attempt to minimize the project duration. Programs vary as to the number of different types of resources that can be used.

34. *Cost and resource curves.* Some programs with cost features will plot curves of budgeted costs, forecasted costs, and/or actual costs as the project proceeds. Other programs will plot a curve of planned costs versus actual costs. Some of the programs with resource features will plot histograms, or profiles, of planned and actual resource usage.

35. *Monte Carlo simulation.* This feature allows the user to designate a separate and distinct frequency distribution (normal, Beta, triangular) for each activity in the network. The user then assigns an optimistic, most likely, and pessimistic time estimate for each activity. With the Monte Carlo simulation procedure, the network calculations are performed many times (100, 1,000, or as many as the user chooses). In each "trial," or pass through the network, an activity duration is randomly chosen for each activity (according to its designated frequency distribution) so that with each pass through the network the length of the most critical path, as well as the activities that comprise the most critical path, may change due to the combination of activity durations which are used for a particular pass, or trial.

The result of the Monte Carlo simulation process is a frequency distribution for the length of the most critical path that allows the user to find the probability for any particular total project duration. Furthermore, a criticality index for each activity is calculated as being the percent of the time that a particular activity occurred on the most critical path out of the total number of Monte Carlo trials, or simulations. Also, a criticality index for each path in the network is calculated as being the percent of the time that a particular path occurred as the most critical path out of the total number of Monte Carlo trials, or simulations.

Since the Monte Carlo process simulates the project many times, each time with a different combination of activity durations, it truly takes into account the uncertainty in estimating activity duration.

There are some computer programs that have only a few of the 35 features listed on the preceding pages, while other, more sophisticated programs may contain many of these features. When deciding on the selection of a computer program as an aid for planning, scheduling, and controlling projects, the user must first identify the minimum features that are needed. This decision can be partly determined by the input data which are readily available (cost estimates, resource estimates, scheduled times) and the kinds of output reports needed (calendar dates, resource profiles, graphical output).

Conclusion

Using a computer does not replace the planning, scheduling, or controlling functions of network planning techniques, nor does it replace the management decision making process. The primary value of the computer is the speed at which it will perform the arithmetic calculations required in developing the schedule and the ability to produce a variety of outputs.

A computer is not always required, and certain factors must be considered when determining whether to use a computer or to make the calculations manually. Even if it is decided to use a computer, there may be several computer programs available for use. These programs may vary from simple and straightforward to highly sophisticated, and the user must then decide which computer program to use, depending on what input data are available and what kinds of output reports are needed. The remaining pages of this chapter show sample computer output reports for the outhouse project example discussed in the previous chapters.

Appendix L provides an alphabetical listing of companies that supply computer programs for project planning, scheduling, and control.

NETWORK PLANNING AND SCHEDULING

CONSTRUCTION OF A PRIMARY SEWAGE TREATMENT FACILITY

USER IDENTIFICATION JAGIDO
MASTER IDENTIFICATION CRAP01
REPORT DATE 01JUN70

RUN NUMBER 01
ACTIVITY REPORT 001
PAGE 1

PROJECT -OUTHOUSE-
SORTED BY PREDECESSOR, SUCCESSOR

PRED. EVENT	SUCC. EVENT	ACTIVITY CODE	ACTIVITY DESCRIPTION	TIME (DAYS)	EARLIEST START	EARLIEST FINISH	LATEST START	LATEST FINISH	TOTAL SLACK	ACTUAL FINISH
000-001	000-002		TEST WIND AND CHOOSE LOCATION	15	15JUN70	03JUL70	02JUN70	22JUN70	- 9	N
000-001	000-003		DECIDE ON STYLE	6	15JUN70	22JUN70	25JUN70	02JUL70	8	N
000-001	000-016		COLLECT OLD NEWSPAPER	10	15JUN70	26JUN70	05AUG70	18AUG70	37	N
000-002	000-004		CLEAR LAND	11	06JUL70	20JUL70	23JUN70	07JUL70	- 9	N
000-003	000-004		DRAW PLANS	3	23JUN70	25JUN70	03JUL70	07JUL70	8	N
000-004	000-005		BUY MATERIALS	5	21JUL70	27JUL70	08JUL70	14JUL70	- 9	N
000-004	000-012		DIG HOLE	13	21JUL70	06AUG70	16JUL70	03AUG70	- 3	N
000-004	000-013		CLEAR PATH FROM HOUSE	7	21JUL70	29JUL70	28JUL70	05AUG70	5	N
000-005	000-006		BUILD FRAME	4	28JUL70	31JUL70	15JUL70	20JUL70	- 9	N
000-005	000-010		BUILD DOOR	1	28JUL70	28JUL70	29JUL70	29JUL70	1	N
000-006	000-007		PUT ON ROOF	1	03AUG70	03AUG70	21JUL70	21JUL70	- 9	N
000-007	000-008		PUT ON EXTERIOR SIDING	2	04AUG70	05AUG70	22JUL70	23JUL70	- 9	N
000-008	000-009		PUT UP INTERIOR WALLS	5	06AUG70	12AUG70	24JUL70	30JUL70	- 9	N
000-009	000-011		DUMMY ACTIVITY	0	12AUG70	12AUG70	30JUL70	30JUL70	- 9	N
000-009	000-012		SAND SEAT	1	13AUG70	13AUG70	03AUG70	03AUG70	- 8	N
000-010	000-011		CUT HALF MOON IN DOOR FOR LIGHT	1	29JUL70	29JUL70	30JUL70	30JUL70	1	N
000-011	000-012		MOUNT DOOR	2	13AUG70	14AUG70	31JUL70	03AUG70	- 9	N
000-012	000-013		POSITION STRUCTURE OVER HOLE	2	17AUG70	18AUG70	04AUG70	05AUG70	- 9	N
000-013	000-014		PAINT INTERIOR	4	19AUG70	24AUG70	06AUG70	11AUG70	- 9	N
000-013	000-016		PAINT EXTERIOR	7	19AUG70	27AUG70	10AUG70	18AUG70	- 7	N
000-014	000-015		INSTALL DOOR LATCH	2	25AUG70	26AUG70	17AUG70	18AUG70	- 6	N
000-014	000-016		HANG PICTURES	5	25AUG70	31AUG70	12AUG70	18AUG70	- 9	N
000-015	000-016		DUMMY ACTIVITY	0	26AUG70	26AUG70	18AUG70	18AUG70	- 6	N
000-016	000-017		TEST	9	01SEP70	14SEP70	19AUG70	31AUG70	- 9	N

NETWORK PLANNING AND SCHEDULING

CONSTRUCTION OF A PRIMARY SEWAGE TREATMENT FACILITY

USER IDENTIFICATION JAGIDO
MASTER IDENTIFICATION CRAPO1
REPORT DATE 01JUN70

RUN NUMBER 01
EVENT REPORT 002
PAGE 1

PROJECT -OUTHOUSE-
SORTED BY EVENT NUMBER

EVENT NUMBER	L C	EVENT TITLE	CRITICAL PRED.	L C	EARLIEST EXPECTED	LATEST ALLOWABLE	EVENT SLACK	SCHEDULED DATE	ACTUAL DATE	
000-001		START PROJECT			15JUN70	02JUN70	- 9	15JUN70		N
000-002			000-001		03JUL70	22JUN70	- 9			
000-003			000-001		22JUN70	02JUL70	8			
000-004			000-002		20JUL70	07JUL70	- 9			
000-005			000-004		27JUL70	14JUL70	- 9			
000-006			000-005		31JUL70	20JUL70	- 9			
000-007			000-006		03AUG70	21JUL70	- 9			
000-008			000-007		05AUG70	23JUL70	- 9			
000-009			000-008		12AUG70	30JUL70	- 9			
000-010			000-005		28JUL70	29JUL70	1			
000-011			000-009		12AUG70	30JUL70	- 9			
000-012			000-011		14AUG70	03AUG70	- 9			
000-013		START PAINTING	000-012		18AUG70	05AUG70	- 9	14AUG70		N
000-014			000-013		24AUG70	11AUG70	- 9			
000-015			000-014		26AUG70	18AUG70	- 6			
000-016			000-014		31AUG70	18AUG70	- 9			
000-017		OUTHOUSE COMPLETED	000-016		14SEP70	31AUG70	- 9	31AUG70		N

NETWORK PLANNING AND SCHEDULING

CONSTRUCTION OF A PRIMARY SEWAGE TREATMENT FACILITY

USER IDENTIFICATION JAGIDO
MASTER IDENTIFICATION CRAP01
REPORT DATE 01JUN70

RUN NUMBER 01
ACTIVITY REPORT 003
PAGE 1

PROJECT -OUTHOUSE-
SORTED BY SLACK

PRED. EVENT	SUCC. EVENT	ACTIVITY CODE	ACTIVITY DESCRIPTION	TIME (DAYS)	EARLIEST START	EARLIEST FINISH	LATEST START	LATEST FINISH	TOTAL SLACK	ACTUAL FINISH
000-001	000-002		TEST WIND AND CHOOSE LOCATION	15	15JUN70	03JUL70	02JUN70	22JUN70	- 9	N
000-002	000-004		CLEAR LAND	11	06JUL70	20JUL70	23JUN70	07JUL70	- 9	N
000-004	000-005		BUY MATERIALS	5	21JUL70	27JUL70	08JUL70	14JUL70	- 9	N
000-005	000-006		BUILD FRAME	4	28JUL70	31JUL70	15JUL70	20JUL70	- 9	N
000-006	000-007		PUT ON ROOF	1	03AUG70	03AUG70	21JUL70	21JUL70	- 9	N
000-007	000-008		PUT ON EXTERIOR SIDING	2	04AUG70	05AUG70	22JUL70	23JUL70	- 9	N
000-008	000-009		PUT UP INTERIOR WALLS	5	06AUG70	12AUG70	24JUL70	30JUL70	- 9	N
000-009	000-011		DUMMY ACTIVITY	0	12AUG70	12AUG70	30JUL70	30JUL70	- 9	N
000-011	000-012		MOUNT DOOR	2	13AUG70	14AUG70	31JUL70	03AUG70	- 9	N
000-012	000-013		POSITION STRUCTURE OVER HOLE	2	17AUG70	18AUG70	04AUG70	05AUG70	- 9	N
000-013	000-014		PAINT INTERIOR	4	19AUG70	24AUG70	06AUG70	11AUG70	- 9	N
000-014	000-016		HANG PICTURES	5	25AUG70	31AUG70	12AUG70	18AUG70	- 9	N
000-016	000-017		TEST	9	01SEP70	14SEP70	19AUG70	31AUG70	- 9	N
000-009	000-012		SAND SEAT	1	13AUG70	13AUG70	03AUG70	03AUG70	- 8	N
000-013	000-016		PAINT EXTERIOR	7	19AUG70	27AUG70	10AUG70	18AUG70	- 7	N
000-014	000-015		INSTALL DOOR LATCH	2	25AUG70	26AUG70	17AUG70	18AUG70	- 6	N
000-015	000-016		DUMMY ACTIVITY	0	26AUG70	26AUG70	18AUG70	18AUG70	- 6	N
000-004	000-012		DIG HOLE	13	21JUL70	06AUG70	16JUL70	03AUG70	- 3	N
000-005	000-010		BUILD DOOR	1	28JUL70	28JUL70	29JUL70	29JUL70	1	N
000-010	000-011		CUT HALF MOON IN DOOR FOR LIGHT	1	29JUL70	29JUL70	30JUL70	30JUL70	1	N
000-004	000-013		CLEAR PATH FROM HOUSE	7	21JUL70	29JUL70	28JUL70	05AUG70	5	N
000-001	000-003		DECIDE ON STYLE	6	15JUN70	22JUN70	25JUN70	02JUL70	8	N
000-003	000-004		DRAW PLANS	3	23JUN70	25JUN70	03JUL70	07JUL70	8	N
000-001	000-016		COLLECT OLD NEWSPAPER	10	15JUN70	26JUN70	05AUG70	18AUG70	37	N

```
                                                                          15HRS.  49 MIN.    02/14/74

                                       NETWORK PLANNING AND SCHEDULING

MASTER FILE REPORT SUMMARY

  USER IDENTIFICATION- JAGIDO     OLD MASTER IDENTIFICATION- CRAP01     NEW MASTER IDENTIFICATION- CRAP01     RUN NUMBER- 01

  NETWORK TITLE- CONSTRUCTION OF A PRIMARY SEWAGE TREATMENT FACILITY

  REPORT DATE- 01JUN70

  NETWORK START DATE-     SCHEDULED  15JUN70       ACTUAL

  NETWORK FINISH DATE-    SCHEDULED  31AUG70     EXPECTED  14SEP70

  MOST CRITICAL SLACK      -9 WORKDAYS

  ACTIVITIES    24

  EVENTS        17

  REPORT OPTIONS-

    MASTER FILE REPORT OPTION   3

    SUMMARY REPORT OPTION       0

    REPORT DATE OPTION          1

    INTERNAL SORT OPTION        0

    TYPE OF RUN OPTION          0

    WORKDAYS PER WEEK           5

    WORKHOURS PER DAY           8

END OF RUN
```

NETWORK PLANNING AND SCHEDULING

CONSTRUCTION OF A PRIMARY SEWAGE TREATMENT FACILITY

USER IDENTIFICATION JAGIDO
MASTER IDENTIFICATION CRAPO2
REPORT DATE 23JUL70

RUN NUMBER 02
ACTIVITY REPORT 001
PAGE 1

PROJECT -OUTHOUSE-
SORTED BY PREDECESSOR, SUCCESSOR

PRED. EVENT	SUCC. EVENT	ACTIVITY CODE	ACTIVITY DESCRIPTION	TIME (DAYS)	EARLIEST START	EARLIEST FINISH	LATEST START	LATEST FINISH	TOTAL SLACK	ACTUAL FINISH	
000-001	000-002		TEST WIND AND CHOOSE LOCATION							02JUL70	A
000-001	000-003		DECIDE ON STYLE							08JUL70	A
000-001	000-016		COLLECT OLD NEWSPAPER	10	*	* |23JUL70|	11AUG70	24AUG70	22		
000-002	000-004		CLEAR LAND							13JUL70	A
000-003	000-004		DRAW PLANS							16JUL70	A
000-004	000-005		BUY MATERIALS							21JUL70	A
000-004	000-012		DIG HOLE							23JUL70	A
000-004	000-013		CLEAR PATH FROM HOUSE	9	*17JUL70*	29JUL70	30JUL70	11AUG70	9		T
000-005	000-006		BUILD FRAME	3	*22JUL70*	24JUL70	*23JUL70*	27JUL70	1		T
000-005	000-010		BUILD DOOR							22JUL70	A
000-006	000-007		PUT ON ROOF	1	27JUL70	27JUL70	28JUL70	28JUL70	1		
000-007	000-008		PUT ON EXTERIOR SIDING	2	28JUL70	29JUL70	30JUL70	31JUL70	2		
000-007	000-018		INSTALL VENT PIPE	2	28JUL70	29JUL70	29JUL70	30JUL70	1		N
000-008	000-009		PUT UP INTERIOR WALLS	3	31JUL70	04AUG70	03AUG70	05AUG70	1		T
000-009	000-011		DUMMY ACTIVITY	0	04AUG70	04AUG70	05AUG70	05AUG70	1		
000-009	000-012		SAND SEAT	1	05AUG70	05AUG70	07AUG70	07AUG70	2		
000-010	000-011		CUT HALF MOON IN DOOR FOR LIGHT	1	*	* |23JUL70|	05AUG70	05AUG70	9		
000-011	000-012		MOUNT DOOR	2	05AUG70	06AUG70	06AUG70	07AUG70	1		
000-012	000-013		POSITION STRUCTURE OVER HOLE	2	07AUG70	10AUG70	10AUG70	11AUG70	1		
000-013	000-014		PAINT INTERIOR	4	11AUG70	14AUG70	12AUG70	17AUG70	1		
000-013	000-016		PAINT EXTERIOR	7	11AUG70	19AUG70	14AUG70	24AUG70	3		
000-014	000-015		INSTALL DOOR LATCH	2	17AUG70	18AUG70	21AUG70	24AUG70	4		
000-014	000-016		HANG PICTURES	5	17AUG70	21AUG70	18AUG70	24AUG70	1		
000-015	000-016		DUMMY ACTIVITY	0	18AUG70	18AUG70	24AUG70	24AUG70	4		
000-016	000-017		TEST	9	24AUG70	03SEP70	25AUG70	04SEP70	1		
000-018	000-008		SHINGLE ROOF	1	30JUL70	30JUL70	31JUL70	31JUL70	1		N

NETWORK PLANNING AND SCHEDULING

CONSTRUCTION OF A PRIMARY SEWAGE TREATMENT FACILITY

USER IDENTIFICATION JAGIDD
MASTER IDENTIFICATION CRAP02
REPORT DATE 23JUL70

RUN NUMBER 02
EVENT REPORT 002
PAGE 1

PROJECT -OUTHOUSE-
SORTED BY EVENT NUMBER

EVENT NUMBER	L C	EVENT TITLE	CRITICAL PRED.	L C	EARLIEST EXPECTED	LATEST ALLOWABLE	EVENT SLACK	SCHEDULED DATE	ACTUAL DATE
000-001		START PROJECT						15JUN70	18JUN70 A
000-002									02JUL70 A
000-003									08JUL70 A
000-004									16JUL70 A
000-005									21JUL70 A
000-006			000-005		24JUL70	27JUL70	1		
000-007			000-006		27JUL70	28JUL70	1		
000-008			000-018		30JUL70	31JUL70	1		
000-009			000-008		04AUG70	05AUG70	1		
000-010									22JUL70 A
000-011			000-009		04AUG70	05AUG70	1		
000-012			000-011		06AUG70	07AUG70	1		
000-013		START PAINTING	000-012		10AUG70	11AUG70	1	14AUG70	
000-014			000-013		14AUG70	17AUG70	1		
000-015			000-014		18AUG70	24AUG70	4		
000-016			000-014		21AUG70	24AUG70	1		
000-017		OUTHOUSE COMPLETED	000-016		03SEP70	04SEP70	1	04SEP70	S
000-018			000-007		29JUL70	30JUL70	1		

NETWORK PLANNING AND SCHEDULING

CONSTRUCTION OF A PRIMARY SEWAGE TREATMENT FACILITY

USER IDENTIFICATION JAGIDO
MASTER IDENTIFICATION CRAP02
REPORT DATE 23JUL70

RUN NUMBER 02
ACTIVITY REPORT 003
PAGE 1

PROJECT -OUTHOUSE-
SORTED BY SLACK

PRED. EVENT	SUCC. EVENT	ACTIVITY CODE	ACTIVITY DESCRIPTION	TIME (DAYS)	EARLIEST START	EARLIEST FINISH	LATEST START	LATEST FINISH	TOTAL SLACK	ACTUAL FINISH	
000-001	000-002		TEST WIND AND CHOOSE LOCATION							02JUL70	A
000-001	000-003		DECIDE ON STYLE							08JUL70	A
000-003	000-004		DRAW PLANS							16JUL70	A
000-002	000-004		CLEAR LAND							13JUL70	A
000-004	000-005		BUY MATERIALS							21JUL70	A
000-005	000-006		BUILD FRAME	3	*22JUL70*	24JUL70	*23JUL70*	27JUL70	1		T
000-006	000-007		PUT ON ROOF	1	27JUL70	27JUL70	28JUL70	28JUL70	1		
000-018	000-008		SHINGLE ROOF	1	30JUL70	30JUL70	31JUL70	31JUL70	1		N
000-008	000-009		PUT UP INTERIOR WALLS	3	31JUL70	04AUG70	03AUG70	05AUG70	1		T
000-009	000-011		DUMMY ACTIVITY	0	04AUG70	04AUG70	05AUG70	05AUG70	1		
000-011	000-012		MOUNT DOOR	2	05AUG70	06AUG70	06AUG70	07AUG70	1		
000-012	000-013		POSITION STRUCTURE OVER HOLE	2	07AUG70	10AUG70	10AUG70	11AUG70	1		
000-013	000-014		PAINT INTERIOR	4	11AUG70	14AUG70	12AUG70	17AUG70	1		
000-014	000-016		HANG PICTURES	5	17AUG70	21AUG70	18AUG70	24AUG70	1		
000-016	000-017		TEST	9	24AUG70	03SEP70	25AUG70	04SEP70	1		
000-007	000-018		INSTALL VENT PIPE	2	28JUL70	29JUL70	29JUL70	30JUL70	1		N
000-007	000-008		PUT ON EXTERIOR SIDING	2	28JUL70	29JUL70	30JUL70	31JUL70	2		
000-009	000-012		SAND SEAT	1	05AUG70	05AUG70	07AUG70	07AUG70	2		
000-013	000-016		PAINT EXTERIOR	7	11AUG70	19AUG70	14AUG70	24AUG70	3		
000-014	000-015		INSTALL DOOR LATCH	2	17AUG70	18AUG70	21AUG70	24AUG70	4		
000-015	000-016		DUMMY ACTIVITY	0	18AUG70	18AUG70	24AUG70	24AUG70	4		
000-005	000-010		BUILD DOOR							22JUL70	A
000-010	000-011		CUT HALF MOON IN DOOR FOR LIGHT	1	* *	|23JUL70|	05AUG70	05AUG70	9		
000-004	000-013		CLEAR PATH FROM HOUSE	9	*17JUL70*	29JUL70	30JUL70	11AUG70	9		T
000-004	000-012		DIG HOLE							23JUL70	A
000-001	000-016		COLLECT OLD NEWSPAPER	10	* *	|23JUL70|	11AUG70	24AUG70	22		

```
                                                                              9HRS.  34 MIN.   02/15/74

                                              NETWORK PLANNING AND SCHEDULING

MASTER FILE REPORT SUMMARY

  USER IDENTIFICATION- JAGIDO     OLD MASTER IDENTIFICATION- CRAP01     NEW MASTER IDENTIFICATION- CRAP02     RUN NUMBER- 02

  NETWORK TITLE- CONSTRUCTION OF A PRIMARY SEWAGE TREATMENT FACILITY

  REPORT DATE- 23JUL70

  NETWORK START DATE-    SCHEDULED  15JUN70       ACTUAL  18JUN70

  NETWORK FINISH DATE-   SCHEDULED  04SEP70     EXPECTED  03SEP70

  MOST CRITICAL SLACK       1 WORKDAYS

  ACTIVITIES    26

  EVENTS        18

  REPORT OPTIONS-

    MASTER FILE REPORT OPTION   3

    SUMMARY REPORT OPTION       0

    REPORT DATE OPTION          1

    INTERNAL SORT OPTION        0

    TYPE OF RUN OPTION          0

    WORKDAYS PER WEEK           5

    WORKHOURS PER DAY           8

END OF RUN
```

CHAPTER 8

Summary and Conclusions

Network planning is a quantitative management technique used for planning, scheduling, and controlling projects that consist of many interrelated activities, but it is not restricted to use by management personnel. It can be used profitably by anyone managing a program or project. It does not tell a person how to plan or manage a project; it helps the users to plan their own method of accomplishing the project and control the progress of the project.

The planning function involves determining the project objectives, then preparing a list of activities necessary to accomplish the objectives, and, finally, developing a network diagram to graphically portray the logical precedence relationships among the activities necessary to accomplish the project objectives.

The scheduling function begins with making estimates of the duration of each activity in the network. Next, scheduled times are chosen for the network start event(s) and network finish event(s). Finally, a schedule is calculated to show the earliest expected and latest allowable start and finish times for each activity in the network. The scheduling function is portrayed in Figure 36 on page 104.

The controlling function involves analyzing the schedule to determine the project status and what replanning may be needed to keep it on schedule. Also, information is gathered regarding the actual times for completed portions of the project. Finally, the plan and schedule for the remaining uncompleted portions of the project are revised and updated to show the effects of both replanning and actual progress.

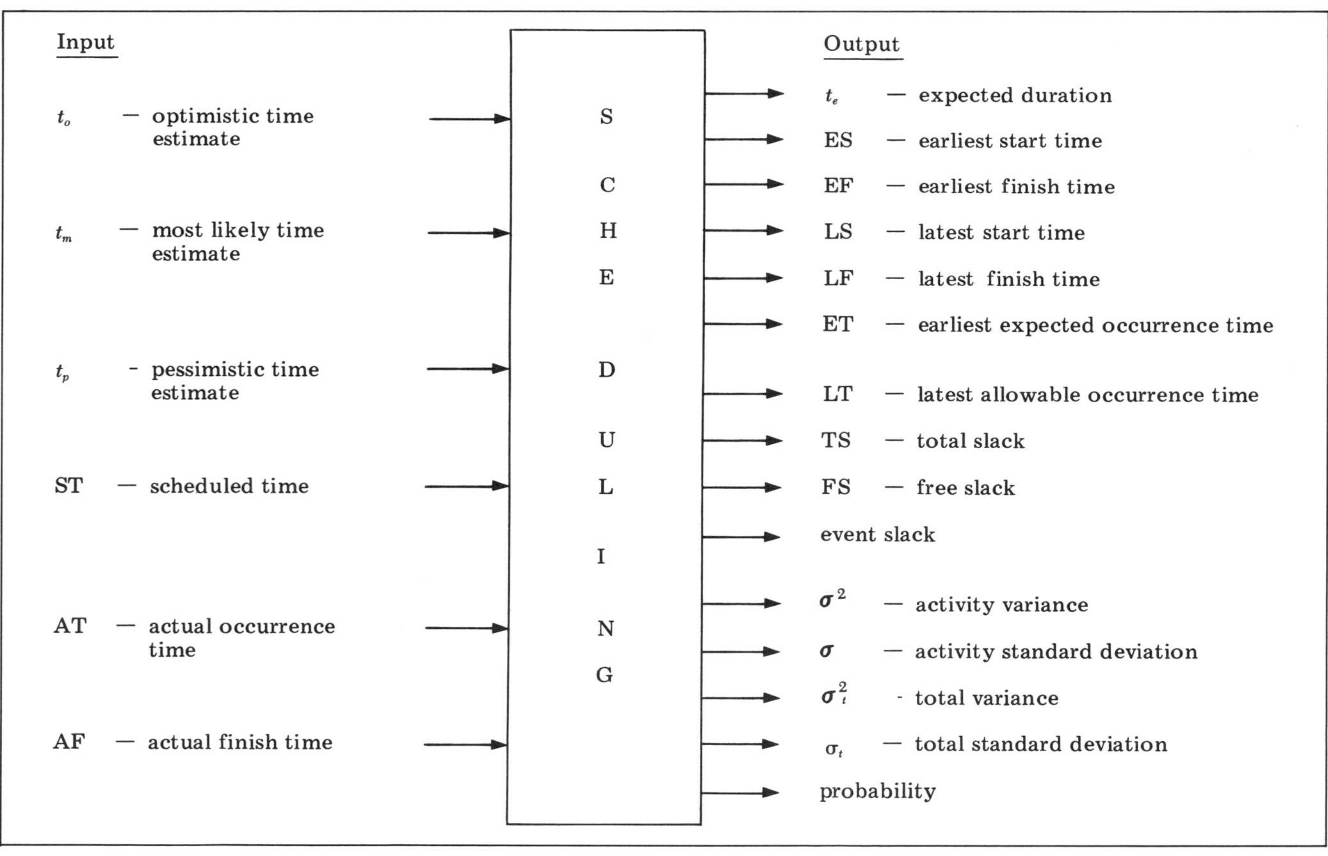
Input
t_o — optimistic time estimate
t_m — most likely time estimate
t_p - pessimistic time estimate
ST — scheduled time
AT — actual occurrence time
AF — actual finish time
S C H E D U L I N G
Output
t_e — expected duration
ES — earliest start time
EF — earliest finish time
LS — latest start time
LF — latest finish time
ET — earliest expected occurrence time
LT — latest allowable occurrence time
TS — total slack
FS — free slack
event slack
σ^2 — activity variance
σ — activity standard deviation
σ_t^2 - total variance
σ_t — total standard deviation
probability

Figure 36.

Network planning is most often compared to the bar chart method, a more traditional planning and scheduling technique. The planning and scheduling functions of a bar chart must be considered simultaneously, since the bar chart is drawn to a time scale. On the other hand, network planning allows for a separation of these functions, since the network diagram is not drawn to any time scale. The network diagram also gives a visual representation of the interrelationships of activities, while in the bar chart the tasks are considered somewhat independently. The Gantt, or bar, chart indicates only "forward" calculations (equivalent to ES and EF times) and does not consider "backward" calculations (equivalent to LS and LF times). So, one may be inclined to schedule overtime or add resources before there is a real need, since slack cannot be calculated by the bar chart technique. The bar chart method can be considered a static technique, since the bar chart is cumbersome to revise and update as the project progresses, while network planning can be considered a dynamic technique, since it allows for a continuous monitoring and controlling of the progress of a project.

Network planning is a systems approach to planning, since it integrates all portions or subsystems of a project into one total system and enables the user to view the plan and schedule for the entire system or project.

Network planning is not a panacea. It is merely a project management tool or aid, and it will not replace the decision making process. Rather, it allows for "management by exception" by highlighting those portions of a project that require problem solving and decision making. It is essential that the implementation of network planning be given management support at all levels in an organization, and that the proper use of the technique be understood by those people directly responsible for achieving the project objective.

Advantages and Difficulties

Once network planning has been used on a project and its effectiveness reviewed upon completion of a project, some of its advantages and difficulties become clear. Network planning can be used to various degrees on a project: it can be used only to develop an initial plan; it can be extended to also develop an initial schedule; or it can be fully utilized to control the project progress.

One advantage of network planning is that it provides a master plan for a project. Any project needs some type of plan, and a network plan seems to work better than most other techniques, since it shows the interrelationships of all portions of a project. One of the most important benefits of network planning comes from developing the network diagram, since this forces the user to completely "think through" the entire project and to plan on a systematic and logical basis. Often, while developing a network diagram for a project, alternate approaches to accomplishing the project may come into focus which were not obvious or thought of previously. Also, the network diagram provides an easily understood visual or graphical portrayal of the interdependence of the activities necessary to accomplish project objectives.

Another advantage of network planning is that it takes uncertainty into account by permitting the use of three estimates for the duration of an activity.

Network planning allows the user to simulate alternate plans or courses of action when developing the initial plan and when replanning during the project. It can be used to provide a sensitivity analysis for these alternate courses of action by simulating changes in the plan and then analyzing the schedule to evaluate the effect of the changes on the schedule.

Many project planning and scheduling computer programs are available for faster and more accurate data handling and schedule calculations. The network schedules provide a vehicle for reporting on actual, expected and allowable progress, and project status by supplying information on completed, in-progress, and unstarted activities.

A great advantage of network planning is its ability to forecast. It not only predicts when each activity is expected to start and finish, but also when each activity must start and finish in order to accomplish the project objective on its scheduled time. It also points out the areas of the project that are behind schedule or may be in potential trouble. This allows the project manager to employ "management by exception," and direct his or her attention to critical activities. It also provides a quantitative basis for making decisions for replanning and controlling the project to alleviate these critical areas. This replaces the use of "seat of the pants" decision making, which is often used in attempting to control a project when network planning techniques are not used.

Although network planning is basically a time-oriented technique, it helps in the planning of resources. This can be accomplished by simulating alternate plans or by checking the earliest and latest start times to see when various resources will be needed. Furthermore, network planning helps in reallocating resources as the project progresses by pointing out the areas where the application of resources would be most effective.

Probably one of the best benefits of network planning is that it provides a basis for communication among people involved in accomplishing the project. It provides a communication link which should allow for a smoother flow of information and result in better coordination among all people involved.

One limitation of using network planning is the possible difficulty of obtaining accurate and reliable input information. For example, a major misuse of network planning is that of using unrealistic duration estimates. If you try to give shorter estimates of activity duration than are realistic, you fool no one but yourself, and you will soon find that you are missing the earliest finish times and are falling behind schedule as the project progresses. It must be emphasized that network planning is only as good as the input data provided.

Another difficulty arises in getting people to understand the schedule reports. Unless the implications of a network schedule are completely understood, there is a possibility that the schedule reports will be misinterpreted and result in improper replanning and wrong decision making. Network planning is often misused by people who use the schedule strictly for documentation purposes only and do not use it as a basis for making decisions for controlling the progress of a project.

Ineffective decision making can result if there is a long time delay between the collection of data on actual progress and the calculation of an associated updated schedule, since this updated schedule will show information that is not as current as it should be.

A drawback to using network planning may be the time needed to develop and refine the network plan and regularly revise and update it during the life of the project. However, the extra effort and time spent in implementing network planning at the beginning of a project is well worth it when the long-term results and benefits are considered.

CHAPTER 9

An Illustration: Major Electrical Supplier Provides Steel Customer with Better Installation and Start-Up Planning

Many factors can cause delays and slippages in large projects, but none is more serious than the failure to comprehensively plan and schedule the installation and start-up phase of the project. Management techniques, such as network planning, make it possible to improve management control during installation and start-up. This comprehensive planning system is beneficial to everyone involved in the installation.

It has been common practice in steel mill construction to use bar charts as a planning tool. But even with the use of bar charts, some large projects have failed to meet their originally planned completion dates by as much as a year or more. With a network planning diagram, it is possible to show all the vital interrelationships among the tasks in a large complex project. Furthermore, it provides a means of indicating problem areas and their consequent effects on the overall project. By bringing attention to specific potential problem areas or delays before they occur, appropriate action may be planned to either correct or compensate for these delays. It should be noted that, like any other management technique, network planning is limited by the reliability of its input data.

The case that follows explains how network planning was implemented in planning and controlling the installation and start-up phase of a highly automated, computer-controlled cold rolling mill electrical drive system for a large steel company.

Initial Planning

The major electrical equipment supplier for the project had certain responsibilities regarding the successful implementation of the electrical system, including substantial technical and supervisory services during the installation and start-up, so the supplier designated a project management team, which included a project manager, site manager, and network planning analyst. The project manager had the overall responsibility for all electrical equipment his company provided from the time the order was received until the time the mill was started and running properly. The site manager was responsible for all activity at the mill site related to installing and testing the equipment.

The goal of the planning function was to develop a network diagram showing the logical precedence relationship of all the activities necessary to properly install and check out the electrical drive system for this mill. The first step in developing the network diagram was to define the end objective for the project, which was the rolling of a coil through the mill that would meet certain previously defined performance criteria.

The next step was to prepare a list of all the activities necessary to meet this objective. The site manager, who had already managed the installation of several similar projects, was the person primarily responsible for preparing this list of activities. This list included both activities directly related to the electrical equipment and tasks that would have to be performed by the electrical contractor, piping contractor, mill builder, and others. The list of activities contained 498 specific tasks, of which 226, or approximately 45%, were activities performed by the electrical equipment supplier's personnel; the remaining tasks were the responsibility of other contractors.

The site manager then developed a network diagram from the list of activities by drawing the sequence of their logical precedence relationships. Although drawing a network is not an easy task, the extra time and effort to develop a well thought out master plan early in the project are well spent. The site manager, drawing on his experience and discussions with customer personnel and other contractors, was able to develop a network diagram that was to be the master plan for the electrical installation portion of the project. The customer and other contractors were called upon to provide information required during the initial planning stages to provide for a broad and well-coordinated planning effort throughout the project.

Scheduling the Plan

When using network planning, the functions of planning and scheduling can be considered separately. Once the network diagram was drawn, the initial planning function was complete.

The scheduling function began with estimating the expected durations of each of the activities in the network. The site manager made the duration estimates for each of the tasks for which his personnel were responsible, while estimates for the remaining activities were solicited from the proper contractors.

The network diagram showing all the activities with their estimated durations provided more information for planning and scheduling than the conventional bar chart technique, primarily because the network diagram showed the interrelationships of activities to each other, while the bar chart treats each activity independently of other activities. Although it is possible to recognize some of the interdependencies within a group of activities, it is almost impossible to know all of the constraints that activities have on each other in a large steel mill installation unless the network planning approach is used.

The electrical equipment supplier's installation personnel arrived on the site sometime after the overall project had begun. The work of the electrical and mechanical contractors was already in progress, so a commonly used network diagram, having just one network start event, could not be used. Instead a multi-start network had to be used, but with this multi-start network, it was necessary to estimate a scheduled start date for each network start event in the network. There were a total of 71 network start events, and the scheduled start dates for each were estimated by the site manager in conjunction with the customer.

In addition to estimates of the duration for each activity and the scheduled dates for the network start events, the scheduled finish date for completion of the project objective was needed. This date was designated by the customer.

Computer-Generated Schedule Reports

Having the network diagram, the estimated duration for each activity, the estimated scheduled dates for each network start event, and the project objective, the next step was to calculate the earliest finish date, latest finish date, and slack for each activity in the network. It was necessary to use a computer to make these schedule calculations because of the size and complexity of the network.

The network logic and all the data were input to a project scheduling computer program. This program checked for errors in the input data and for errors in network logic, such as loops. If the input data were free of errors, the computer then made the proper calculations and printed the beginning event number, ending event number, responsibility code, activity description, earliest finish date, latest finish date, expected activity duration, slack, and actual completion date (for completed activities only) for each activity in the network. There were also flags, printed out to call attention to those activities that should have been in-progress and those that were overdue for completion as of the report date that the schedule was issued.

Although the computer program allowed the schedule to be sorted in various ways, it was decided that the most useful reports would be those which sorted the schedule for the activities in the following three ways:

1. Major sort by latest finish date, subsort by slack;
2. Major sort by responsibility code, first level subsort by latest finish date, second level subsort by slack; and
3. Major sort by slack, subsort by latest finish date.

All three reports contained the same information; it was just a matter of its being sorted differently. Copies of these reports were distributed as follows:

Electrical equipment supplier (three)

- Project manager
- Site manager
- District field engineering office

Customer (three)

- Engineering department
- Construction department

Planning department

Electrical contractor (one)

The customer engineering department was responsible for all the engineering throughout the project, including both mechanical and electrical equipment. The construction department had the overall responsibility for the construction and installation phase of the project. The planning department was responsible for the overall coordination and for reporting to management on the status of the project.

Controlling the Project

Once the initial plan and schedule had been developed, a system for controlling the project to keep it on schedule was developed. There were bi-weekly meetings at which schedule reports were discussed, new information was considered, and decisions regarding replanning were made. The site manager, customer personnel, and representatives of the various contractors attended these meetings, which took place at the mill site. At these meetings data regarding the actual completion dates of activities, actual dates of network start events, any changes in activity durations or scheduled dates, and the addition, deletion, or rearrangement of activities were reported. Schedules were analyzed to find the most critical areas in the project and to make sure that work was being performed on the critical activities rather than on activities with large amounts of positive slack. In other words, the computer reports or schedules pointed out the areas where slippages had occurred and were expected to occur in the schedule, and based on this information, the customer and all the contractors were able to make necessary planning decisions to bring the project back on schedule. Also, the reasons for slippage in the schedule were discussed. The primary reasons seemed to be: material not being delivered on time, bad weather conditions during the winter months, and engineering changes made on equipment already installed or manufactured. Finally, the status of all activities that should have been in-progress and which were expected to be completed by the next meeting were discussed.

The data generated at these meetings were input to the computer for an updated set of schedule reports which were issued to the recipients previously mentioned. This two-week updating cycle continued throughout the project and was an effective method of controlling project progress.

Conclusion

The project was completed two weeks ahead of the customer's required date. Starting up the steel mill two weeks earlier than originally scheduled can be looked upon as a major achievement, considering the highly complex interdependencies of all installation contractors for this multi-million dollar project and considering the history of past steel mill start-ups. Not only did this mill start up on time, but it was done without any major unresolved problems with the electrical equipment. This, too, can be looked upon as a major achievement, since the electrical equipment included much automation and highly sophisticated controls, including two computers.

Appendices

A—Network diagram (activity-on-the-arrow format)
B—Network diagram (activity-in-the-box format)
C—Network diagram with schedule input data (activity-on-the-arrow format)
D—Network diagram with schedule input data (activity-in-the-box format)
E—Updated and revised network diagram (activity-on-the-arrow format)
F—Updated and revised network diagram (activity-in-the-box format)
G—Answers to questions within Chapters 2, 3, 4, and 5
H—Answers to questions at the end of Chapters 2, 3, 4, and 5
I—Nomenclature
J—Glossary
K—Recommended references
L—Project management software suppliers

APPENDIX A

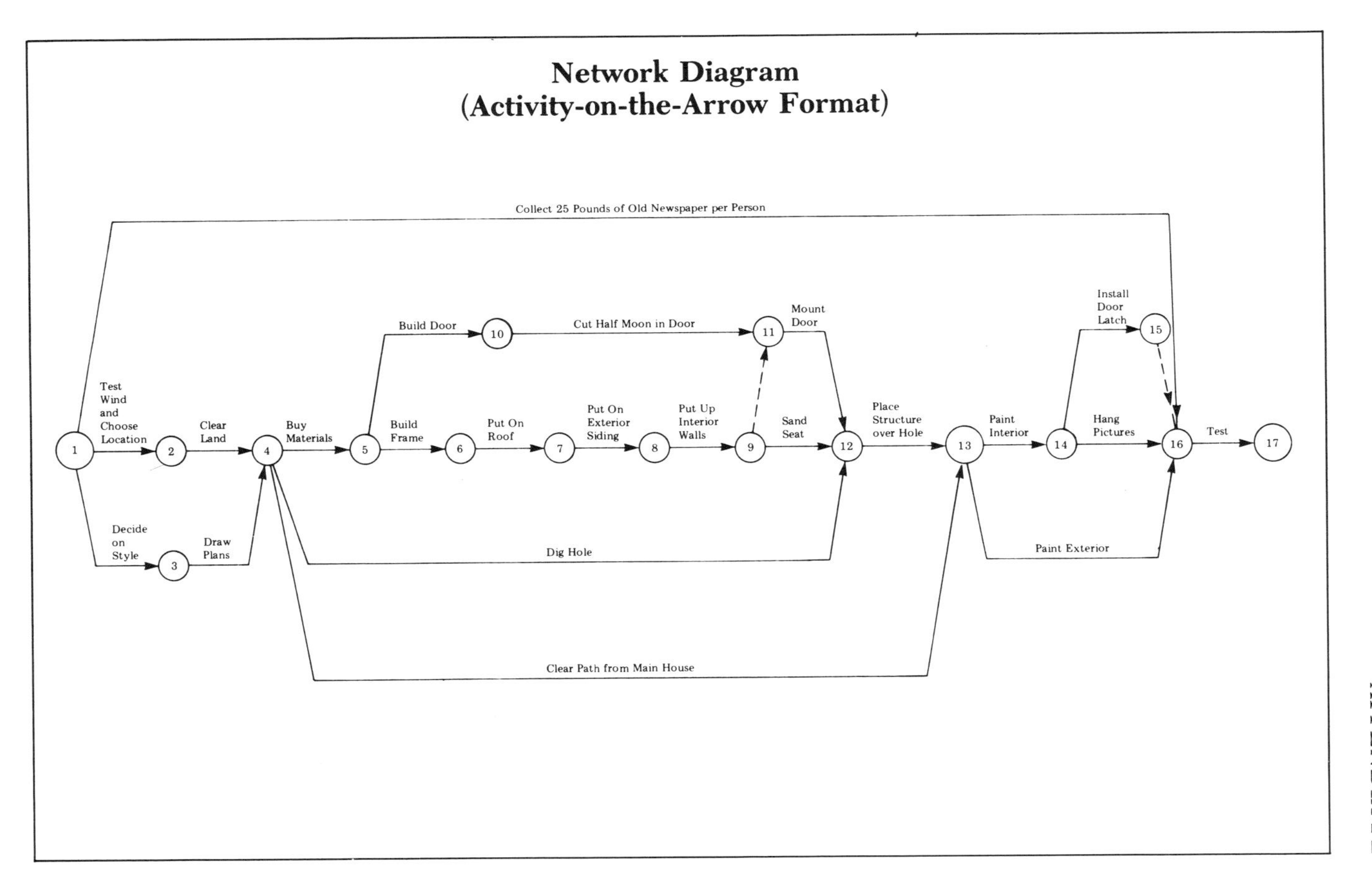
Network Diagram
(Activity-on-the-Arrow Format)
Collect 25 Pounds of Old Newspaper per Person
Build Door
Cut Half Moon in Door
Mount Door
Install Door Latch
Test Wind and Choose Location
Clear Land
Buy Materials
Build Frame
Put On Roof
Put On Exterior Siding
Put Up Interior Walls
Sand Seat
Place Structure over Hole
Paint Interior
Hang Pictures
Test
Decide on Style
Draw Plans
Dig Hole
Paint Exterior
Clear Path from Main House

APPENDIX B

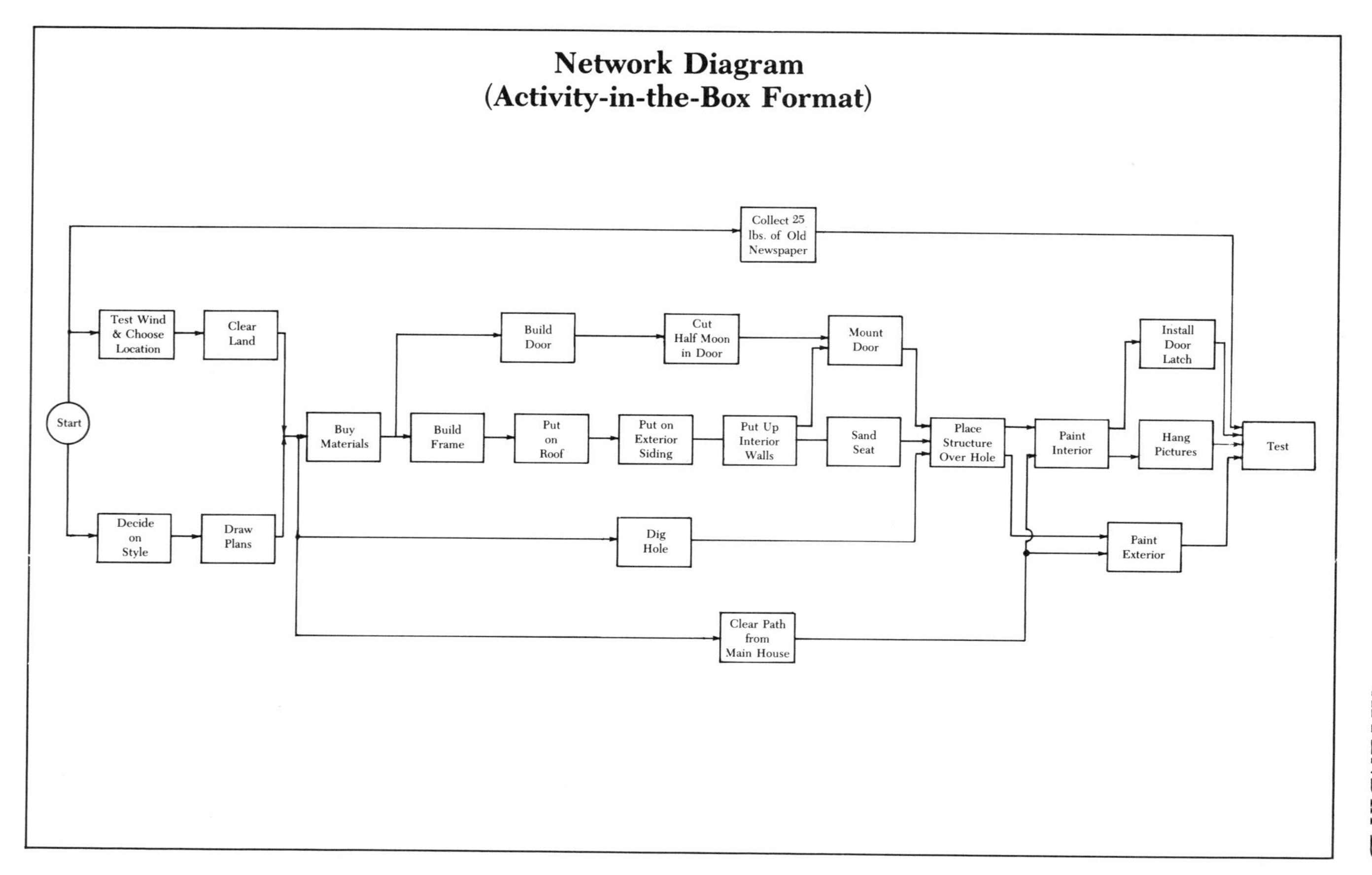

Network Diagram (Activity-in-the-Box Format)

APPENDIX C

Network Diagram with Schedule Input Data (Activity-on-the-Arrow Format)

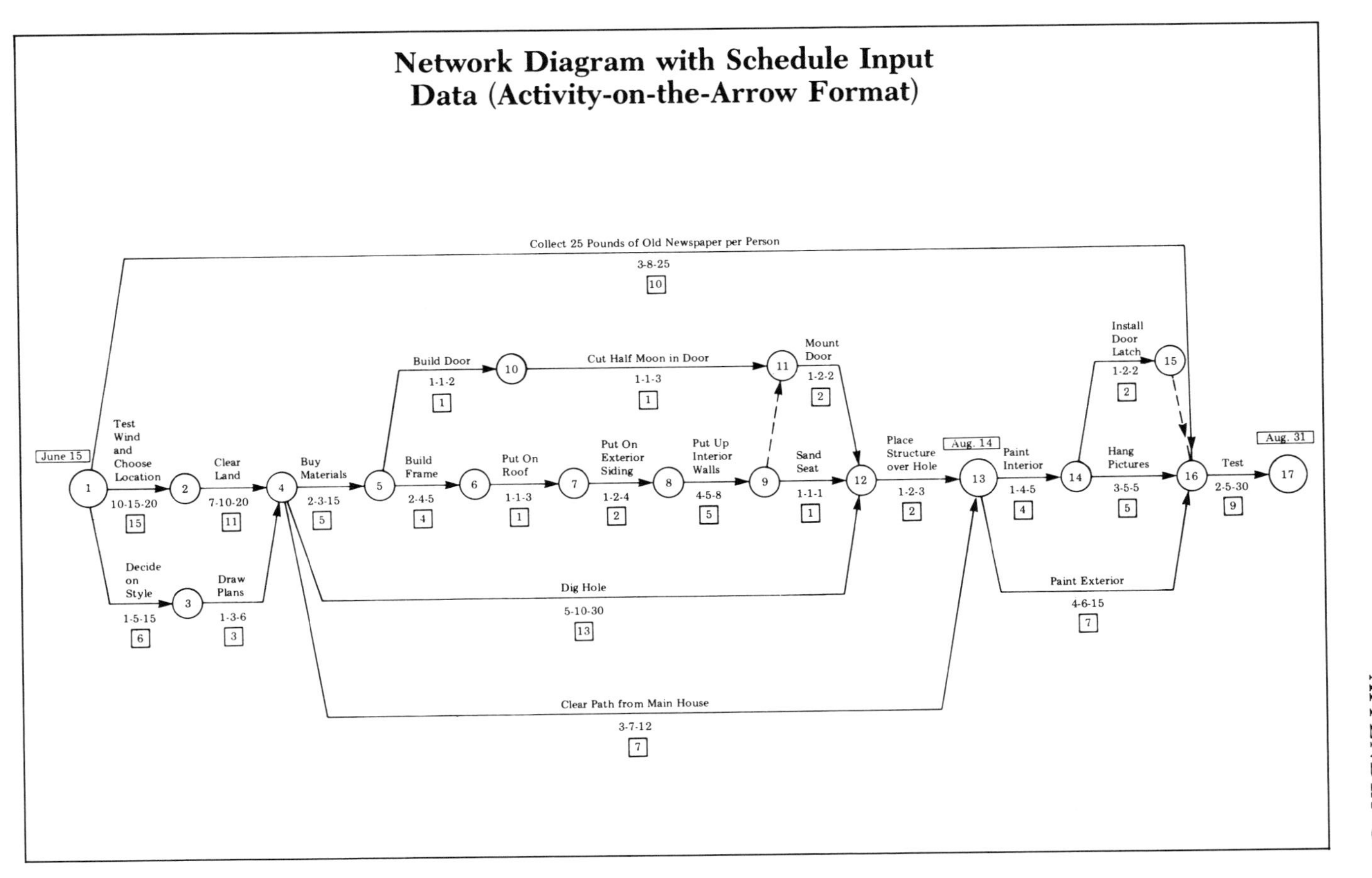

APPENDIX D

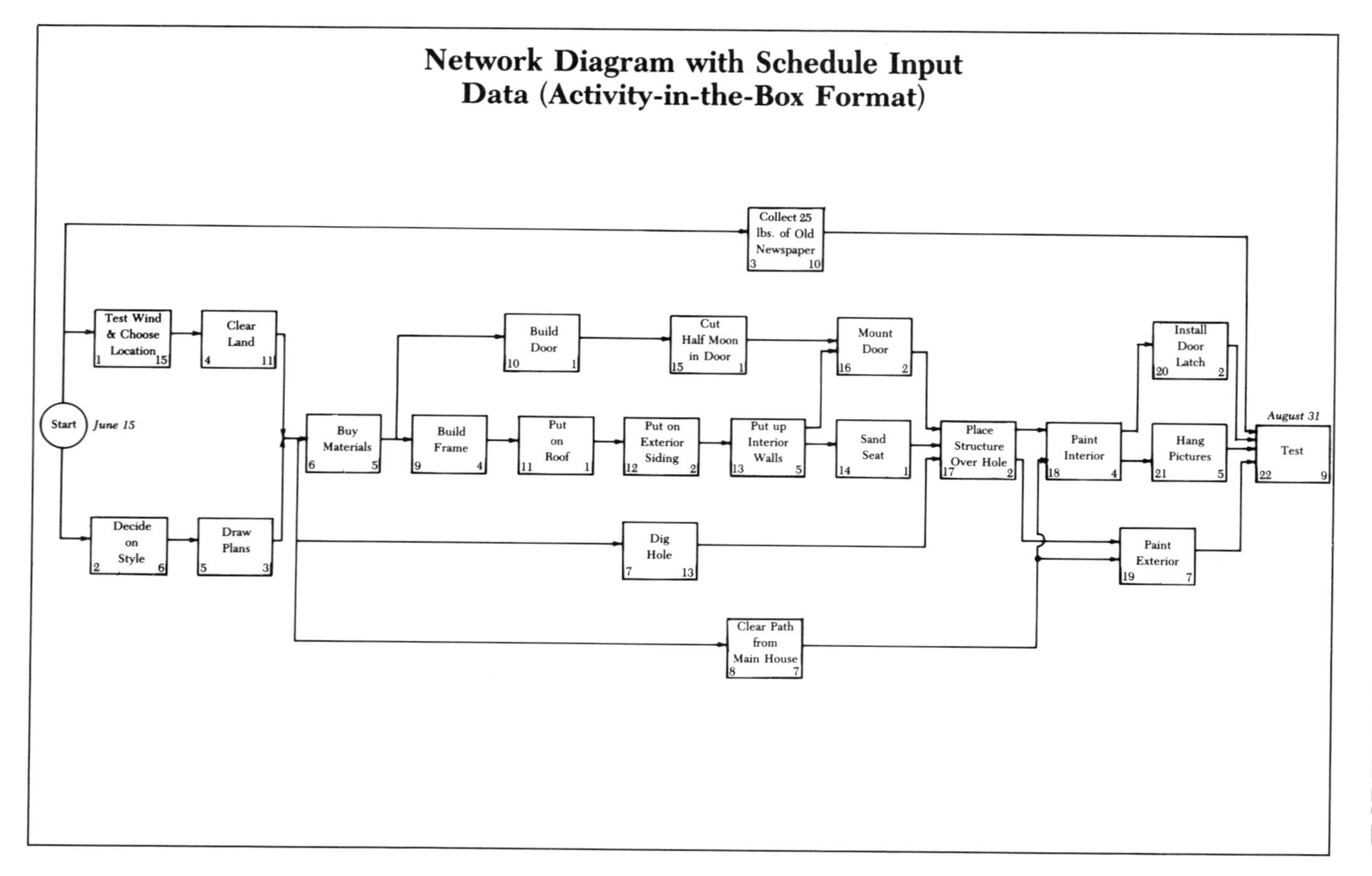
Network Diagram with Schedule Input Data (Activity-in-the-Box Format)
Start
June 15
Test Wind & Choose Location 1 15
Clear Land 4 11
Decide on Style 2 6
Draw Plans 5 3
Collect 25 lbs. of Old Newspaper 3 10
Buy Materials 6 5
Build Frame 9 4
Build Door 10 1
Put on Roof 11 1
Put on Exterior Siding 12 2
Put up Interior Walls 13 5
Cut Half Moon in Door 15 1
Sand Seat 14 1
Mount Door 16 2
Dig Hole 7 13
Clear Path from Main House 8 7
Place Structure Over Hole 17 2
Paint Interior 18 4
Install Door Latch 20 2
Hang Pictures 21 5
Paint Exterior 19 7
August 31
Test 22 9

APPENDIX E

Updated and Revised Network Diagram (Activity-on-the-Arrow Format)

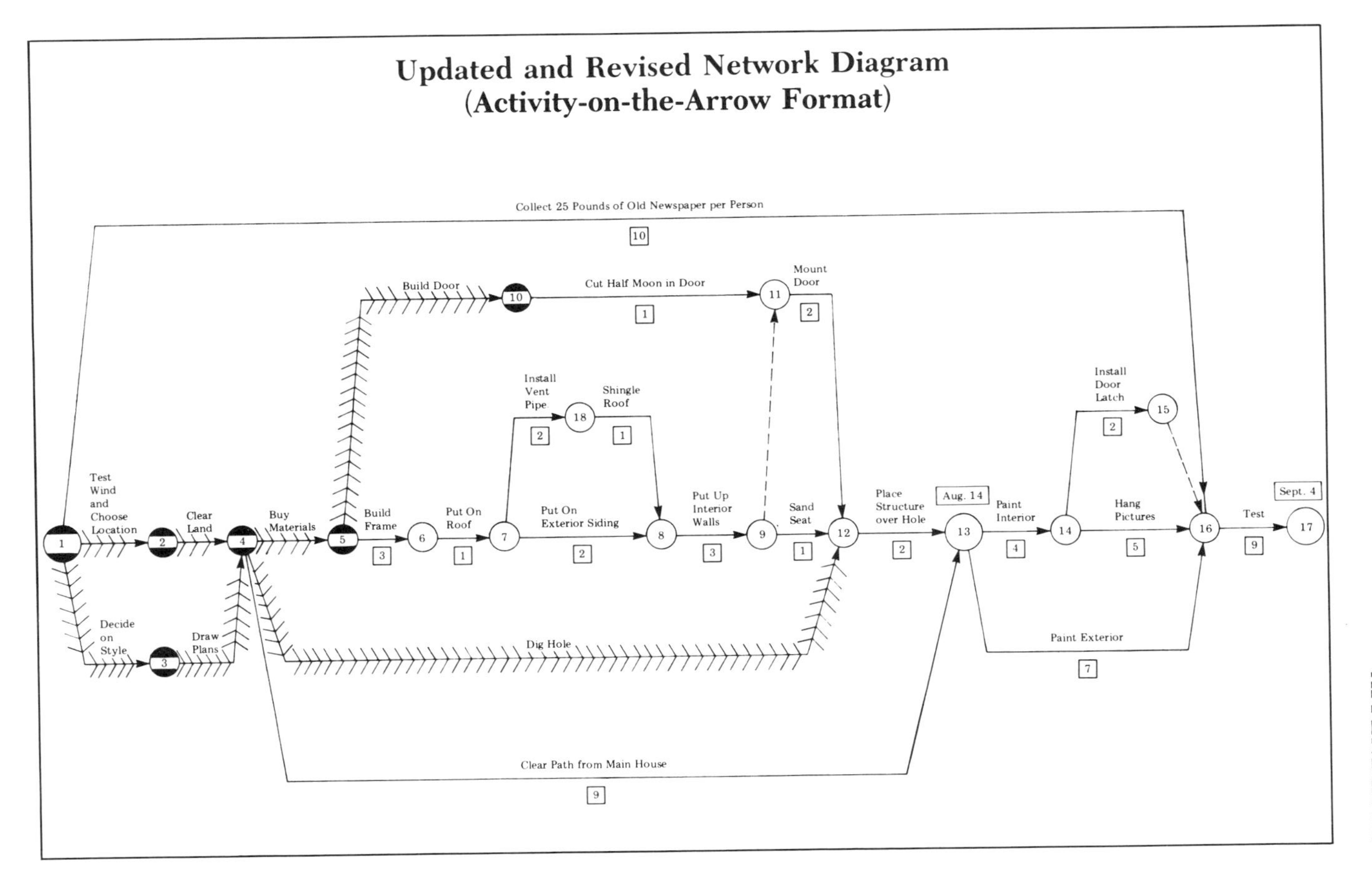

APPENDIX F

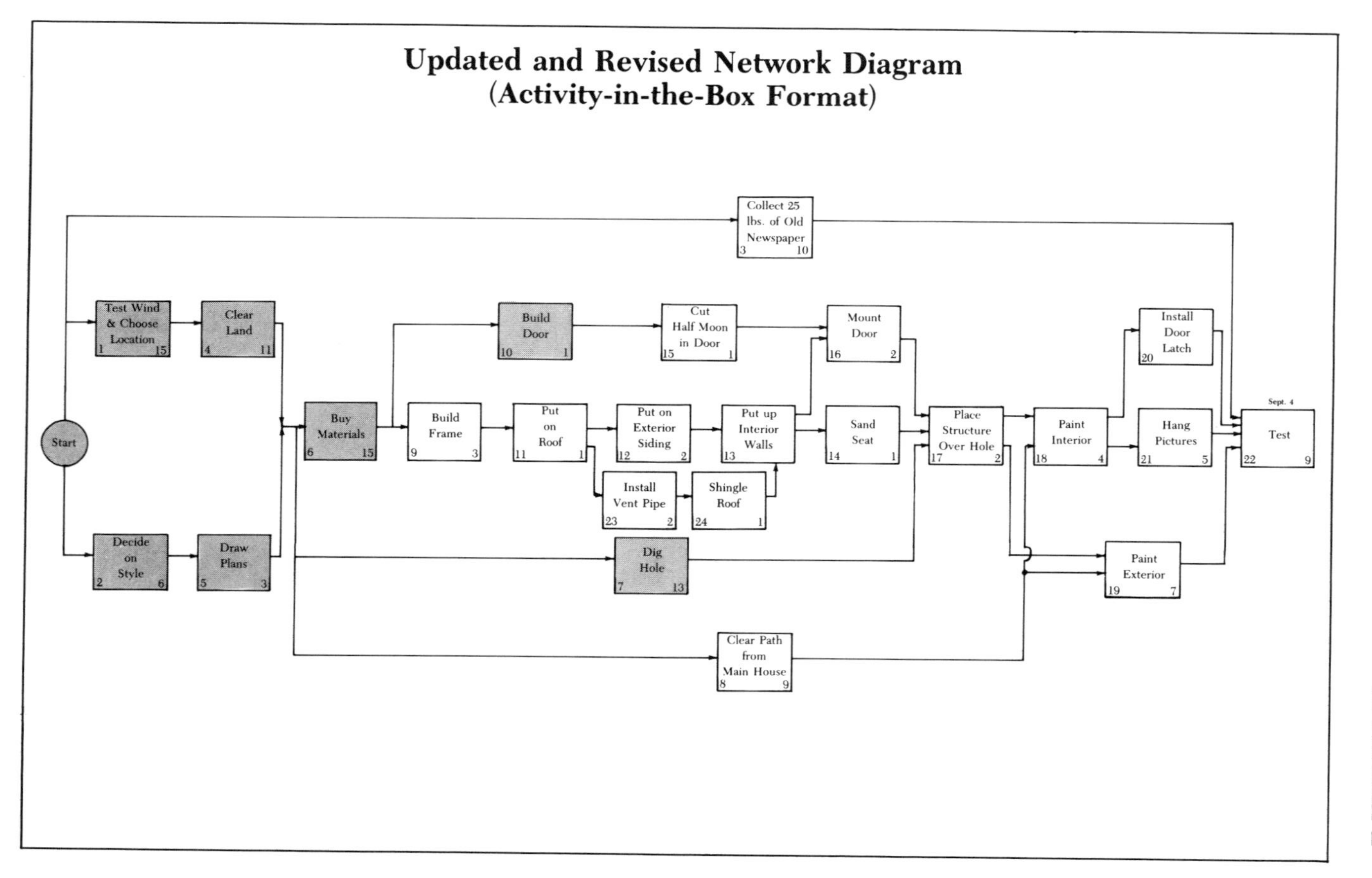

Updated and Revised Network Diagram
(Activity-in-the-Box Format)
Start
Test Wind & Choose Location 1 15
Clear Land 4 11
Decide on Style 2 6
Draw Plans 5 3
Buy Materials 6 15
Collect 25 lbs. of Old Newspaper 3 10
Build Door 10 1
Cut Half Moon in Door 15 1
Mount Door 16 2
Build Frame 9 3
Put on Roof 11 1
Put on Exterior Siding 12 2
Put up Interior Walls 13
Install Vent Pipe 23 2
Shingle Roof 24 1
Sand Seat 14 1
Dig Hole 7 13
Clear Path from Main House 8 9
Place Structure Over Hole 17 2
Paint Interior 18 4
Paint Exterior 19 7
Install Door Latch 20
Hang Pictures 21 5
Sept. 4
Test 22 9

APPENDIX G

Answers to Questions within Chapters 2, 3, 4, and 5

Chapter 2

A. 1.
a. Defining the project objectives;
b. Preparing a list of activities necessary to accomplish the project objectives;
c. Developing a network diagram.

A. 2. An **activity** is the expenditure of effort over some *period of time*; whereas an **event** is the start or finish of an activity and is an instant or *point in time*.

A. 3.

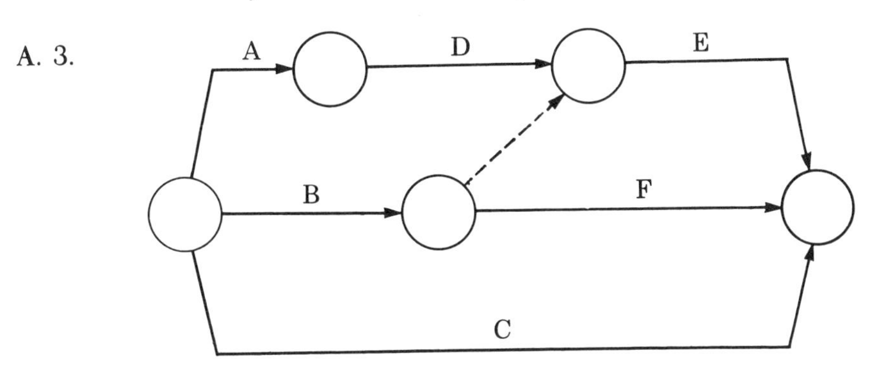

A. 4. A summary network is one which is made up of a small number of generalized or high level activities rather than a large number of detailed activities.

A. 5.
a. The structure may be placed over the hole.
b. Paint the interior; install the door latch; hang pictures; collect newspaper.
c. Build frame.
d. 6
e. 17

Chapter 3

A. 6. (1)-b, (2)-c, (3)-a.

A. 7. 1-1-3; zero.

A. 8. The three time estimates allow the user to take into account "uncertainty" when estimating the duration for an activity.

A. 9. August 31.

A. 10.

$$t_e = \frac{8 + 4(12) + 22}{6} = 13$$

A. 11. a.

$$t_e = \frac{1 + 4(1) + 7}{6} = 2$$

b. 50% chance that it will actually take less than t_e, and a 50% chance that it will actually take longer.

A. 12. July 27; August 6; July 29.

A. 13. July 18; July 17

A. 14. LF for both 4-13 and 12-13 is August 5.
LS for 4-13 is July 28; and LS for 12-13 is August 4.

A. 15.
a. July 30.
b. September 24.

A. 16.

EVENT	ET	LT
1	June 15	June 2
5	July 27	July 14
7	August 3	July 21
12	August 14	August 3
16	August 31	August 18
17	September 14	August 31

A. 17. Before the scheduled time of the network finish event.

A. 18. Less than 50–50. (Remember for network finish events ST = LT.)

A. 19. a.

ACTIVITY	FREE SLACK
A	0
B	8
C	12
D	20
E	33

b. Event slack for event 99 is -12.

Chapter 4

A. 20. Expected duration: $$t_e = \frac{5 + 4(8) + 23}{6} = 10.$$

Variance: $$\sigma^2 = \left(\frac{23 - 5}{6}\right)^2 = 9.$$

A. 21. 34%.

A. 22. $\sigma^2 = 25$.

Since there are a total of four standard deviations ($+2$ and -2) between 12 and 32, then $4\sigma = 32 - 12 = 20$ and thus $1 = 5$. Therefore, variance $= \sigma^2 = 5 \times 5 = 25$.

Chapter 5

A. 23. a. To compare actual progress to the schedule.
b. If actual progress is behind schedule, the network can be replanned.

A. 24. July 8; June 25; July 1.

A. 25. Those which have the least value of total slack.

APPENDIX H

Answers to Questions at End of Chapters 2, 3, 4, and 5

Chapter 2

1. a. Define project objectives.
   b. Prepare a list of activities necessary to accomplish project objectives.
   c. Develop a network diagram.

2. a. **Activity**—An expenditure of effort over some period of time on a particular part of a project.
   b. **Event**—The start or finish of an activity. It consumes no time, and therefore is an instant or point in time. Events link activities together.
   c. **Loop**—A path of activities which goes in a continuous circle. This is not allowed when drawing a network diagram.
   d. **Dummy**—An activity that consumes no time. It is represented by a dashed arrow in a network diagram.
   e. **Network start event**—An event that has no activities entering into it, and has only activities leading from it.
   f. **Network finish event**—An event which has only activities entering into it and has no activities leading from it.

3. (1) g
   (2) h
   (3) a
   (4) k
   (5) f
   (6) m
   (7) c

4.

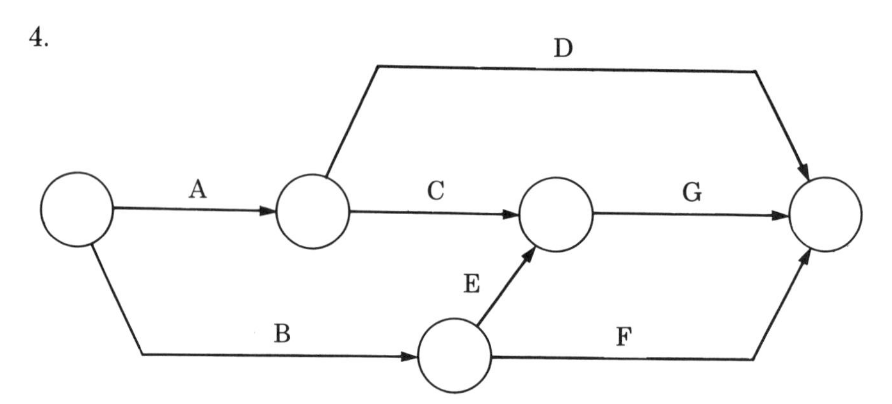

5.

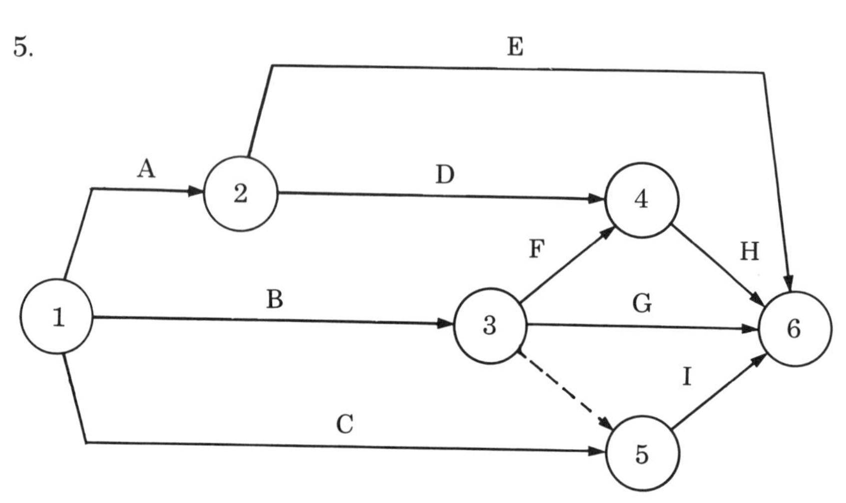

6. a. Duplicate event number 3.
   b. Activities A and D have the same predecessor-successor event number combination, 1-2.
   c. Activities B and E are not necessary.
   d. Activity J has two arrowheads.
   e. Activity G has no arrowhead.
   f. Activity N has no ending event.
   g. Activities C-K-H form a loop.

Chapter 3

1. a. False
   b. False
   c. False
   d. False
   e. True
   f. True
   g. False
   h. False
   i. True
   j. False

2. a. **Optimistic time estimate** (t_o)—The time in which a particular activity may be completed if everything goes well and there are no complications.
   b. **Earliest start time (ES)**—The earliest time a particular activity is expected to begin, based on calculations made from the scheduled time of the network start event and from the expected durations of the preceding activities.
   c. **Latest finish time (LF)**—The latest time by which a particular activity must be completed in order not to jeopardize the completion of the project by the scheduled time of the network finish event. This time is based on calculations made from the scheduled time of the network finish event and from the expected durations of succeeding activities.
   d. **Earliest expected occurrence time (ET)**—The earliest time a particular event is expected to occur. For a given event, it is equal to the latest time of all the earliest finish times of all the activities entering into that event.
   e. **Total slack (TS)**—The maximum amount of time that the activities on a particular path can be delayed without jeopardizing the completion of the project by the scheduled time of the network finish event. For a particular activity: TS = LF − EF = LS − ES. The total slack for a particular path is shared by all the activities on that path.

3. (1) f
   (2) d
   (3) m
   (4) l
   (5) h
   (6) i
   (7) c
   (8) k
   (9) e
   (10) a
   (11) b
   (12) j
   (13) g

4.

Start Event	Finish Event		t_e	ES	EF	LS	LF	TS
1	2	A	10	6/1	6/12	5/28	6/10	−2
1	3	B	5	6/1	6/5	5/29	6/4	−1
2	4	C	6	6/15	6/22	6/11	6/18	−2
2	5	E	3	6/15	6/17	6/18	6/22	3
3	4	F	10	6/8	6/19	6/5	6/18	−1
3	5	D	11	6/8	6/22	6/8	6/22	0
4	6	G	8	6/23	7/2	6/19	6/30	−2
5	6	H	6	6/23	6/30	6/23	6/30	0

Chapter 4

1. True

2.
$$t_e = \frac{t_o + 4(t_m) + t_p}{6} = \frac{2 + 56 + 14}{6} = \frac{72}{6} = 12$$

$$\sigma^2 = \left(\frac{t_p - t_o}{6}\right)^2 = \left(\frac{14 - 2}{6}\right)^2 = \left(\frac{12}{6}\right)^2 = (2)^2 = 4$$

$$\sigma = \sqrt{\sigma^2} = \sqrt{4} = 2$$

3. b

4.
$$Z = \frac{ST - ET}{\sigma_t} = \frac{5}{4} = 1.25$$

From the table on page 57, for Z = 1.25, the area between ST and ET is 0.39435. Therefore, the probability of finishing before ET = .50000 + .39435 = .89435 or 89.435%.

Chapter 5

1. a. In order to determine if actual progress is ahead or behind of the schedule, and
   b. To replan the remaining portion of the project if actual progress is behind schedule.

2. a. AT is the actual occurrence time for an event.
   b. AF is the actual finish time for an activity.

3. a. Earliest start time (ES).
   b. Earliest finish time (EF).
   c. Earliest expected time to occur (ET).
   d. Slack (TS).
   e. Probability of meeting scheduled times.

4. True

5. a. If possible, perform activities in parallel instead of in series.
   b. Reduce the technical specifications or requirements.
   c. Use additional resources.
   d. Use automated equipment.
   e. Work overtime.

APPENDIX I

Nomenclature

AF	Actual finish time of an activity
AT	Actual occurrence time of an event
CPM	Critical Path Method
EF	Earliest expected finish time for an activity
ES	Earliest expected start time for an activity
ET	Earliest expected occurrence time for an event
FS	Free slack
LF	Latest allowable finish time for an activity
LS	Latest allowable start time for an activity
LT	Latest allowable occurrence time for an event
NFE	Network finish event
NSE	Network start event
PDM	Precedence Diagramming Method
PERT	Program Evaluation and Review Technique
RT	Report time
ST	Scheduled time
TS	Total slack

t_e	Expected duration of an activity
t_m	Most likely duration estimate for an activity
t_o	Optimistic duration estimate for an activity
t_p	Pessimistic duration estimate for an activity
WBS	Work Breakdown Structure
σ	Standard deviation
σ^2	Variance

APPENDIX J

Glossary

Activity—The expenditure of effort over some period of time on a particular part of a project.

Actual Finish Time (AF)—The time at which a completed activity was actually finished.

Actual Time (AT)—The time at which an event actually occurred.

Bar Chart—A project planning and scheduling technique in which each task is considered independent and is graphically portrayed by a bar, the length of which depends on the expected duration of the activity.

Control—To direct the progress of a plan.

CPM—Critical Path Method. A network planning technique.

Critical Path—Any path in the network that has a negative value of total slack.

Dummy Activity—An activity that does not consume time and has an expected duration of zero. It is represented by a dashed arrow in a network diagram.

Duration—The time, including both applied time and waiting or stand-by time, needed to perform an activity from start to finish.

Earliest (Expected) Finish Time (EF)—The earliest time a particular activity is expected to be completed. It is equal to the activity's earliest expected start time plus the activity's expected duration: EF = ES + t_e.

Earliest (Expected) Start Time (ES)—The earliest time a particular activity is expected to begin, based on calculations made from the scheduled time of the network start event and from the expected durations of preceding activities.

Earliest Expected Time (ET)—The earliest time a particular event is expected to occur. For a given event it is equal to the latest time of all the earliest finish times of all the activities entering into that event.

Event—The start or finish of an activity. Since it does not consume time, it is an instant or point in time. Events link activities together.

Event Slack—The maximum amount of time that the occurrence of a particular event may be delayed without jeopardizing the completion of the project by the scheduled time for the network finish event. For a given event: Event Slack = LT − ET. This is also equal to the least value of total slack of all the activities entering into that given event.

Expected Duration (t_e)—The expected duration for a particular activity. The formula for the three time estimates is:

$$t_e = \frac{t_o + 4(t_m) + t_p}{6}.$$

There is a 50–50 chance that the actual duration will be more or less than t_e.

Float—*See* Slack.

Free Slack (FS)—The amount of time that a particular activity can be delayed without delaying the earliest start time of any immediately succeeding activity. It is the relative difference in the amount of total slack between the activity with the least value of total slack entering into a given event and the values of total slack of all the activities entering into that same event. It is never a negative value.

Laddering—A method for showing the logical precedence relationship of a set of several serial activities that is repeated several consecutive times.

Latest (Allowable) Finish Time (LF)—The latest time by which a particular activity must be completed to ensure the completion of the project by the scheduled time of the network finish event. This time is based on calculations made from the scheduled time of the network finish event and from the expected durations of succeeding activities.

Latest (Allowable) Start Time (LS)—The latest time by which a particular activity must begin so that the project can be completed by the scheduled time of the network finish event. It is equal to the activity's latest allowable finish time minus the activity's expected duration. $LS = LF - t_e$.

Latest Allowable Time (LT)—The latest time at which a particular event must occur to ensure the completion of the project by the scheduled time for the network finish event. For a given event, it is equal to the latest finish time of all the activities entering into that event.

Loop—A path of activities that goes in a continuous circle. This is not allowed in a network diagram.

Management by Exception—Managing a project by directing primary attention to the critical areas of the project.

Mean—A measure of the central tendency of a probability distribution. It divides the area under the probability distribution curve into two equal parts. It is also called the expected value.

Milestone Event—A key or important event, other than a network start event or network finish event, in the network diagram. It may or may not be accompanied by a scheduled time.

Most Critical Path—The most time-consuming path through a network. Also the path which has the least value of total slack.

Most Likely Time (t_m)—The time in which a particular activity can most often be completed under normal conditions.

Multiple Finish Network—A network that has more than one network finish event.

Multiple Start Network—A network that has more than one network start event.

Network—*See* Network diagram.

Network Analysis—Investigating all parts of a schedule to determine the status of the project and its individual activities, the causes of changes in the schedule, and the most effective ways of replanning.

Network Diagram—A visual or graphical representation of a project plan showing all the logical precedence relationships among the activities necessary to achieve the project objectives.

Network Finish Event—An event that has only activities entering into it and does not have activities leading from it. It must be accompanied by a scheduled time.

Network Planning—A general name applied to any project planning and scheduling technique that employs the use of a network diagram.

Network Start Event—An event that does not have activities entering into it and has only activities leading from it. It must be accompanied by a scheduled time.

Node—*See* Event.

Noncritical Path—Any path in the network that has a positive value of total slack.

Optimistic Time (t_o)—The time in which a particular activity may be completed if everything goes well and there are no complications.

Parallel—Concurrently or simultaneously.

Path—A set or chain of serial activities leading from any one event to another event in a network diagram.

PDM—Precedence Diagramming Method. A network planning technique.

PERT—Program Evaluation and Review Technique. A network planning technique.

Pessimistic Time (t_p)—The time in which a particular activity may be completed under an adverse situation, such as having unusual and unforeseen complications.

Plan—A sequence of activities designed to accomplish a set of project objectives.

Precedence—The relationship of certain activities having to be finished before other activities can be started.

Predecessor Event—An event at the start of an activity in a network diagram. Also called beginning event or start event.

Probability—The likelihood of occurrence. It is used with scheduled times for network finish events and milestone events.

Project Objective—The purpose or final goal for undertaking the project. It should be specific and measurable.

Replan—To revise or change the project plan based on an analysis of the schedule.

Report Time (RT)—The time or date when a schedule is calculated.

Resource Allocation—A resource scheduling method which attempts to minimize the project duration with a fixed number of resources.

Resource Leveling (Smoothing)—A resource scheduling method that attempts to minimize both overtime and waiting, or standby, time for resources within a fixed project duration.

Schedule—A timetable for a project plan.

Scheduled Time (ST)—The time chosen for the earliest expected occurrence of a network start event, or the latest allowable occurrence of a network finish event or a milestone event.

Serial—One after another; in tandem.

Slack—An indication of the status of the project and of individual activities. *See* Total slack, Free slack, and Event slack.

Standard Deviation—A measure of the dispersion or spread of a probability distribution. It is equal to the square root of the variance.

Stochastic—Takes into account the probability of occurrence of many values.

Subnetwork—A network for a particular portion or phase of the entire plan.

Successor Event—An event at the end of an activity in a network diagram. Also called ending event or finish event.

Summary Network—A network that contains a small number of generalized, or high-level, activities rather than a large number of detailed activities.

Task—*See* Activity.

Time Estimate—An approximate value for the duration of an activity. It should include both applied time and waiting, or stand-by, time.

Total Slack (TS)—The maximum amount of time that the activities on a particular path can be delayed without jeopardizing the completion of the project by the scheduled time of the network finish event. For a particular activity: TS = LF − EF = LS − ES. The total slack for a particular path is shared by all the activities on that path.

Update—To revise the project plan and schedule based on both replanning and actual progress.

Variance—A measure of the dispersion or spread of a probability distribution. The variance for the particular activity is:

$$\sigma^2 = \left(\frac{t_p - t_o}{6}\right)^2$$

Work Breakdown Structure (WBS)—A graphic technique that divides and subdivides a project into phases, functions, or areas, similar to an organization chart.

APPENDIX K

Recommended References

Cleland, David I. and William R. King, *Project Management Handbook*. New York: Van Nostrand Reinhold Co., 1983.

Kerzner, Harold, *Project Management—A Systems Approach to Planning, Scheduling, and Controlling*. New York: Van Nostrand Reinhold Co., 1979.

Moder, Joseph J., Cecil R. Phillips, and Edward W. Davis, *Project Management with CPM, PERT and Precedence Diagramming*, 3rd ed. New York: Van Nostrand Reinhold Co., 1983.

Rosenau, Milton D., Jr., *Successful Project Management*. Belmont, CA: Lifetime Learning Publications, 1981.

Stuckenbruck, Linn C. (Ed.), *The Implementation of Project Management: The Professional's Handbook*. Reading, MA: Addison-Wesley, 1981.

Wiest, Jerome D. and Ferdinand K. Levy, *A Management Guide to PERT/CPM*, 2nd ed. Englewood Cliffs, NJ: Prentice-Hall, 1977.

APPENDIX L

Project Management Software Suppliers

L-1

This section is an alphabetical listing of companies that supply computer programs for project planning, scheduling, and control. This list was compiled as follows. A list of companies that supply project management software was compiled from various sources. A questionnaire was sent to each of these companies requesting information for inclusion in this listing. An attempt was made to contact those companies that did not respond by the requested date, and another questionnaire was sent where possible. Companies not responding to the initial or second questionnaire are not included in this Appendix. Therefore, there may be additional companies, not listed here, that supply project management software.

This listing contains the company name, address, and telephone number of the company, and the title of their project management computer program(s). Those titles followed by an asterisk are capable of being run on a personal (micro) computer.

1. **A+ Software, Inc.**
16 Academy Street
Skaneateles, NY 13152
315-685-6918
SCHEDULE-IT *

2. **A-Systems Corp.**
1610 South Main Street
Suite G
Bountiful, UT 84010
801-298-0052
CRITICAL PATH SCHEDULE *

3. **Accura Tech, Inc.**
5422 Chevy Chase
Houston, TX 77056
713-960-9385
TIMETABLE

4. **ADP Network Services**
Project Management Services
125 Jackson Plaza
Ann Arbor, MI 48106
313-769-6800
APECS/8000 Project Evaluation & Control System

5. **AGS Management Systems**
880 First Ave.
King of Prussia, PA 19406
215-265-1550
PAC I
PAC II
PAC III
PAC MICRO *

6. **Aha Inc.**
P.O. Box 8405
147 South River Street
Santa Cruz, CA 95060
408-458-9119
PERTMASTER/PERT PLOT *
PMS II/RMS II *
PRIMAVERA PROJECT PLANNER*

7. **ALPHA Software Corp.**
30 B Street
Burlington, MA 01803
617-229-2924
THE EXECUTIVE PACKAGE *

8. **Apple Computer, Inc.**
20525 Mariani Ave.
Cupertino, CA 95014
408-973-2790
MacPROJECT *

9. **Applied Business Technology Corp.**
76 Laight Street
7th Floor
New York, NY 10013
212-219-8945
PROJECT MANAGER WORKBENCH *

10. **Applied Information Development**
823 Commerce Drive
Oak Brook, IL 60521
312-654-3030
PROJECT CONTROL FACILITY *

11. **Applied Management Methods, Inc.**
201 North Broad Street
Doylestown, PA 18901
215-348-1200
TOPMAN Total Planning & Management

12. **ASA**
355 South End Ave.
Suite 26-B
New York, NY 10280
212-321-2408
ASAPMS Project Management System *

13. **Breakthrough Software**
505 San Marin Drive
Novato, CA 94947
415-898-1919
TIME LINE *

14. **The Bridge Inc.**
199 California Drive
Millbrae, CA 94030
415-697-2730
TRAK

15. **M. Bryce & Associates, Inc.**
1248 Springfield Pike
Cincinnati, OH 45215
513-761-8400
PMC2 Project Management Command & Control

16. **Business Support Services, Inc.**
111 West 57th Street
Suite 1102
New York, NY 10019
212-586-1804
PLANTRAC *

17. **CARA Corp.**
611 East Butterfield Road
Suite 110
Lombard, IL 60148
312-968-8100
EPSILON *

18. **COADE**
Division of International Thomson Information Inc.
8550 Katy Freeway
Suite 122
Houston, TX 77024
713-973-9060
PROJECT PLUS *

19. **Collins & Associates**
187 Flying Mist Isle
Foster City, CA 94404
415-571-6991
CPERT

20. **Computerline, Ltd.**
755 Southern Artery
Quincy, MA 02169
617-773-0001
PLANTRAC *

21. **Construction Information Systems, Inc.**
P.O. Box 484
Mill Valley, CA 94942
415-332-5073
SYSTEM 20/20 *

22. **Contel Corp.**
416 West Fifth Ave.
Naperville, IL 60540
312-355-8188
READINET

23. **Control Data Corp.**
Cybernet Services
8100 34th Ave. South
Bloomington, MN 55440
612-853-7914
PRIMAVERA PROJECT PLANNER *
PRIMAVISION *

24. **Convergent Technologies**
Distributed Systems
2441 Mission College Blvd.
MS 7-071
Santa Clara, CA 95051
408-980-9222
PROJECT PLANNER

25. **Corporate Development Associates**
P.O. Box 2546
Upper Union Street
Schenectady, NY 12309
518-346-0698
DMIN *

26. **CRI, Inc.**
Computer Resources Inc.
5333 Betsy Ross Drive
P.O. Box 58004
Santa Clara, CA 95052
408-980-9898
PROJECT ALERT

27. **Data Lab Corp.**
200 West Monroe
Chicago, IL 60606
312-236-8162
PMS-2000

28. **Datamatics**
330 New Brunswick Ave.
Fords, NJ 08863
201-738-9600
PMS Performance Management System *

29. **Decision Science Software**
P.O. Box 7876
Austin, TX 78713
512-926-4527
EXPERT *

30. **Dekker, Ltd.**
214 East Olive Ave.
Redlands, CA 92373
714-793-7939
DEKKER TRAKKER *

31. **Demi-Software**
62 Nursery Road
Ridgefield, CT 06877
203-431-0864
DEMI-PLAN *

32. **DIAC International**
P.O. Box 2105
Seattle, WA 98111
206-281-8556
DIAC EASY TRAK

33. **Digital Marketing Corp.**
2363 Boulevard Circle
Suite 8
Walnut Creek, CA 94595
415-938-2880
MILESTONE *

34. **Dynamic Solutions**
50 Lytton Ave.
Hartsdale, NY 10530
914-949-6058
PERT6

35. **Earth Data Corp.**
User Support Group
P.O. Box 13168
Richmond, VA 23225
804-231-0300
MICROGANTT *

36. **EDUCOL, Inc.**
P.O. Box 726
San Luis Obispo, CA 93406
805-489-0806
PMS Project Management System *

37. **Emerge Systems**
P.O. Box 3175
Indialantic, FL 32903
305-723-0444
GPERT *

38. **Environmental Services, Inc.**
7831 Glenroy Road #340
Minneapolis, MN 55435
612-831-4646
CRAM Critical Resource Allocation Method

39. **GEISCO**
General Electric Information Services Co.
Mail Code TOB2
401 North Washington Street
Rockville, MD 20850
301-340-5688
PVS Project Visibility System

40. **Gilbert Services, Inc.**
CUE Systems
525 Lancaster Ave.
Reading, PA 19603
215-775-2600
GC CUE

41. **Gnomon, Inc.**
P.O. Box 30169
Cincinnati, OH 45230
513-232-3557
GNOMON-CPM

42. **H&S Software**
Arrasmith Trail
Ames, IA 50010
515-232-2331
SCHEDULE-PRO *

43. **Harte Systems, Inc.**
2625 Butterfield Road
Suite 304-N
Oak Brook, IL 60521
312-884-8630
PCS Project Planning System

44. **Harvard Software, Inc.**
521 Great Road
Littleton, MA 01460
617-486-8431
HARVARD PROJECT MANAGER *
HARVARD TOTAL PROJECT MANAGER *

45. **Hollander Associates**
P.O. Box 2276
Fullerton, CA 92633
714-879-9000
TOPS/SCHEDULE

46. **Honeywell, Inc.**
Application Systems Division K31
P.O. Box 8000
Phoenix, AZ 85066
602-862-3559
PMCS Project Management & Control System

47. **IBM**
Information Systems Group
Department 70-R
12 Water Street
White Plains, NY 10601
914-993-7633
CIPREC
CIPREC/GS

48. **ICARUS Corp.**
11300 Rockville Pike
Rockville, MD 20852
301-881-9350
TSS

49. **Institute for Scientific Analysis, Inc.**
ManageMint Systems Group
36 East Baltimore Pike
Suite 106
Media, PA 19063
215-566-0801
MANAGEMINT *

50. **Institute of Industrial Engineers**
Technical Services
25 Technology Park
Norcross, GA 30092
404-449-0460
PROJECT MANAGEMENT GROUP *

51. **ISSCO**
10505 Sorrento Valley Road
San Diego, CA 92121
619-452-0170
TELLAPLAN

52. **K&H Project Systems, Inc.**
48 Woodport Road
Sparta, NJ 07871
201-729-6142
PREMIS

53. **Management & Computer Services**
79 Great Valley Pkwy.
Malvern, PA 19482
563-1140
PROJECTMACS *

54. **Martin Marietta Data Systems**
Information Technology Division
98 Inverness Drive East
Suite 325
Englewood, CO 80112
303-790-3090
TRMS:PS Technical Requirements Management System: Project Status

55. **MC2 Engineering Software**
8107 SW 72 Ave.
Suite 425-E
P.O. Box 430980
Miami, FL 33143
305-665-0100
M2M Critical Path Project Management *

56. **McAUTO**
K253
P.O. Box 516
St. Louis, MO 63166
314-233-1075
MSCS Management Scheduling & Control System *

57. **Metier Management Systems, Inc.**
5884 Point West Drive
Houston, TX 77036
713-988-9100
ARTEMIS *

58. **Micro-Base Corp.**
521 Windsor Park Drive
Dayton, OH 45459
800-338-3508
BAI*PERT

59. **Microsoft**
10700 Northup Way
Bellevue, WA 98004
206-828-8080
MICROSOFT PROJECT *

60. **MISOL**
Promasys
10317 Lake Creek Pkwy.
Building H-1
Austin, TX 78750
512-258-6778
THE PROJECT MANAGER *

61. **Mitchell Management Systems**
2000 West Park Drive
Westborough, MA 01581
617-366-0800
MAPPS
MINI MAPPS
QUICK PLAN *

62. **Morgan Computing Company, Inc.**
10400 North Central Expressway
Suite 210
Dallas, TX 75231
214-739-5895
PATHFINDER *

63. **National Information Systems**
20370 Town Center Lane
Cupertino, CA 95014
408-257-7700
VUE

64. **Nichols & Company, Inc.**
5839 Green Valley Circle
Culver City, CA 90230
213-670-6400
N5500 Project Management System

65. **North America MICA, Inc.**
5230 Carroll Canyon Road
Suite 110
San Diego, CA 92121
619-458-1327
PMS-II Project Management System *

66. **Omicron**
57 Executive Park
Suite 590
Atlanta, GA 30329
404-325-0770
PLAN/TRAX 3.0 *

67. **P C International**
4400 MacArthur Blvd.
Suite 800
Newport Beach, CA 92660
714-476-1020
EASYTRAK *

68. **Philips Group, Inc.**
P.O. Box 14668
Houston, TX 77221-4668
713-747-4733
CPM CPMEDIT *

69. **Pinnell Engineering**
5331 SW Macadam
Suite 270
Portland, OR 97201
503-243-2246
PMS80 *

70. **P M Programming**
5201 East Highland Road
Milford, MI 48042
313-887-3738
CRITPATH *

71. **PMS Project Management Systems**
P.O. Box 2124
Dothan, AL 36302
205-793-0957
PMS Project Management System *

72. **POC-IT Management Services**
606 Wilshire Blvd.
Suite 606
Santa Monica, CA 90401
213-393-4552
MICRO-MAN Project Control System *

73. **P.P.M.C.S.**
1309 East 132nd Street
Burnsville, MN 55337
612-894-2415
NETCON Network Control System *

74. **Primavera Systems, Inc.**
29 Bala Ave.
Bala Cynwyd, PA 19004
215-667-8600
PRIMAVERA PROJECT PLANNER *

75. **PROFITOOL, Inc.**
1600 Stout Street
Suite 2000
Denver, CO 80202-3135
303-571-1555
CMIS Contractor Management Information System

76. **Project Management International**
3735 NW Glenridge
Corvallis, OR 97330
503-754-3436
APSS Automatic Planning & Scheduling System

77. **Project Software & Development, Inc.**
Product Marketing
14 Story Street
Cambridge, MA 02138
617-661-1444
PROJECT/2

78. **Project Software & Development, Inc.**
Micro Products Division
44 Brattle Street
Cambridge, MA 02138
617-661-1444
QWIKNET *

79. **PROMACON, Inc.**
Two Bala Plaza
Suite 925
Bala Cynwyd, PA 19004
215-667-2678
PROMACON/90

80. **Quality Data Products**
101 Little Soda Road
Carson, WA 98610
509-427-4497
QUICK-TROL

81. **Quantitative Software Management, Inc.**
1057 Waverley Way
McLean, VA 22101
703-790-0055
SLIM Software Life Cycle Management *

82. **Radio Shack**
Computer Merchandising
1500 One Tandy Center
Fort Worth, TX 76134
817-338-2248
PROJECT MANAGER *
PROJECT SCHEDULER *

83. **RECON Systems, Inc.**
P.O. Box 14512
Fort Lauderdale, FL 33302
305-486-0929
TIME PRO

84. **Scitor Corp.**
Commercial Division
256 Gibratar Drive
Building B7
Sunnyvale, CA 94089
408-730-0400
PROJECT SCHEDULER 5000 *

85. **Sheppard Software Company**
4750 Clough Creek Road
Redding, CA 96002
916-222-1553
MICROPERT *

86. **Shirley Software Systems**
1936 Huntington Drive
Suite 208
South Pasadena, CA 91030
818-441-5121
MISTER Management Information System for Time, Expense and Resources

87. **Simple Software Inc.**
2 Pinewood
Irvine, CA 92714
714-857-9179
PROJECTMASTER *

88. **SOFTCORP, Inc.**
2340 State Road 580
Suite 244
Clearwater, FL 33575
813-799-3984
ADVANCED PRO-JECT 6 *

89. **Softpoint Company**
1003 Crest Lane
Carnegie, PA 15106
412-279-4130
SOFTPOINT PROJECTS MANAGEMENT SYSTEM *

90. **SOFTRAK Systems**
1977 West North Temple
Salt Lake City, UT 84122
801-531-8550
MICROTRAK *

91. **Software Techniques, Inc.**
4144 North Central Expressway
Suite 960
Dallas, TX 75231
214-823-2784
STATUS

92. **Sperry Information Systems**
Scientific, Energy and Communications Marketing Dept.
P.O. Box 500
Blue Bell, PA 19424
215-542-3998
OPTIMA 1100

93. **Sungard Information Services**
Two Glenhardie Corporate Center
1285 Drummers Lane
Wayne, PA 19087
215-341-8762
SUNPLAN II

94. **SYS COMP Corp.**
2042 Broadway
Santa Monica, CA 90402
213-829-9707
CMCS Construction Management Control System

95. **Systonetics, Inc.**
801 East Chapman Ave.
Fullerton, CA 92631
714-680-0910
VISION Integrated Project Management System EZPERT

96. **T and B Computing, Inc.**
1100 Eisenhower Place
Ann Arbor, MI 48104
313-973-1900
TRACK 50

97. **Tetrad Computer Applications, Ltd.**
1445 West Georgia Street
Vancouver, British Columbia V6G 2T3
CANADA
604-685-2295
SPRED/2

98. **TEVCO, Inc.**
P.O. Box 22614
Sacramento, CA 95822
916-393-1857
CPSS Critical Path Scheduling System

99. **TRIMAG Systems, Inc.**
512 West Lancaster Ave.
Wayne, PA 18041
215-687-9200
CO$TIME

100. **United Information Services Co.**
Division of Control Data Corp.
9300 Metcalf
Overland Park, KS 66212
913-341-9161
OSCAR On-Line System for Controlling Activities & Resources

101. **WELCOM Software Technology**
1325 South Dairy Ashford
Suite 445
Houston, TX 77077
713-558-0514
OPEN PLAN *

102. **WESTICO**
25 Van Zant Street
Norwalk, CT 06855
203-853-6880
MICROGANTT *

103. **Westminster Software**
660 Hansen Way
Suite 2
Palo Alto, CA 94304
415-424-8300
PERTMASTER *

104. **WEXCO International Corp.**
WEXPRO Software Systems
1900 Avenue of the Stars
Suite 1900
Los Angeles, CA 90067
213-556-0547
WEXPRO CONSTRUCTION MANAGEMENT & PROJECT CONTROLS SYSTEM *

105. **Glenn L. White**
10600 Howerton Ave.
Fairfax, VA 22030
703-385-1210
CPMIS

106. **XEBEK, Inc.**
627 Washington Building
Seattle, WA 98101
206-625-0412
PVS Project Visibility System

L-2

This section is an alphabetical listing of the names of computer programs and the reference number (as shown in Appendix L-1) of the supplier of the program.

Program Name	Reference Number
Advanced PROJECT 6	88
APECS/8000	4
APSS	76
ARTEMIS	57
ASAPMS	12
BAI*PERT	58
CIPREC	47
CIPREC/GS	47
CMCS	94
CMIS	75
CO$TIME	99
CPERT	19
CPM CPMEDIT	68
CPMIS	105
CPSS	98
CRAM	38
CRITICAL PATH SCHEDULE	2
CRITPATH	70
DEKKER TRAKKER	30
DEMI-PLAN	31
DIAC EASY TRAK	32
DMIN	25
EASY TRAK	67
EPSILON	17
EXPERT	29
EZPERT	95
GNOMON-CPM	41
GC CUE	40
GPERT	37
HARVARD PROJECT MANAGER	44
HARVARD TOTAL PROJECT MANAGER	44
MacPROJECT	8
MANAGEMINT	49
MAPPS	61
MICROGANTT	35; 102
MICRO-MAN	72
MICROPERT	85
MICROTRAK	90
MICROSOFT PROJECT	59
MILESTONE	33
Mini MAPPS	61
MISTER	86
MSCS	56
M2M	55
NETCON	73
N5500 Project Management System	64
OPEN PLAN	101
OPTIMA 1100	92
OSCAR	100
PAC I	5
PAC II	5
PAC III	5
PAC MICRO	5
PATHFINDER	62
PCS	43
PERTMASTER	6; 103
PERT6	34
PLANTRAC	16; 20
PLAN/TRAK 3.0	66
PMC2	15
PMCS	46
PMS II	6; 65
PMS	28; 36; 71

Program Name	Reference Number
PMS80	69
PMS-2000	27
PREMIS	52
PRIMAVERA PROJECT PLANNER	6; 23; 74
PRIMAVISION	23
PROJECT ALERT	26
Project Control Facility	10
PROJECTMACS	53
Project Management Group	50
PROJECT MANAGER	82
PROJECT MANAGER WORKBENCH	9
PROJECTMASTER	87
PROJECT PLANNER	24
PROJECT PLUS	18
PROJECT SCHEDULER	82
PROJECT SCHEDULER 5000	84
PROJECT/2	77
PROMACON/90	79
PVS	39; 106
QUICK PLAN	61
QUICK-TROL	80
QWIKNET	78

Program Name	Reference Number
READINET	22
SCHEDULE-IT	1
SCHEDULE-PRO	42
SLIM	81
SOFTPOINT Projects Management System	89
SPRED/2	97
STATUS	91
SUNPLAN II	93
SYSTEM 20/20	21
TELLAPLAN	51
The Executive Package	7
The Project Manager	60
TIME LINE	13
TIME PRO	83
TIMETABLE	3
TOPMAN	11
TOPS/SCHEDULE	45
TRACK 50	96
TRAK	14
TRMS:PS	54
TSS	48
VISION	95
VUE	63
WEXPRO Construction Management and Project Control System	104

Index

Typography, cover photograph and design by Steven A. Baron